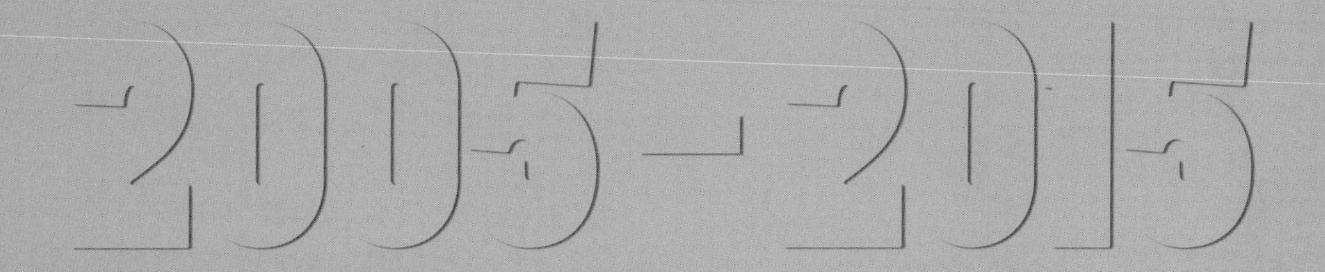

同济都市建筑十年

Selections from Design Works of Urban Architectural Design Institute

Tongji Architectural Design (Group) Co. Ltd.

同济大学建筑设计院（集团）有限公司都市建筑设计院 著

同济大学出版社
TONGJI UNIVERSITY PRESS

封面题字

戴复东

中国工程院院士

都市建筑设计院顾问

Inscription on the Cover Page

DAI Fudong

Member of the Chinese Academy of Engineering

Consultant, Tongji Urban Architectural Design Institute

序

同济大学都市建筑设计院的全称是同济大学建筑设计研究院（集团）有限公司都市建筑设计院，正式成立至今已经整整十年，10 年正是一所建筑设计院开始成熟的年龄，其实这是一所曾经脱胎换骨的设计院，前身可追溯到 1956 年成立的同济大学土木建筑设计院，成立的目的就是为建筑系的教师和学生提供设计实践的场所。我们今天所生活的世界在本质上是经过人工改造的世界，充满了有意识的创造，其中，建筑师功不可没。由于建筑师的特殊专业，建筑师负有重托和信赖，要为失误、疏忽等承担责任，要为社会和人民的生命财产、为人们的生活品质、为社会发展负责，其责任十分重大。历史上建筑师的培养就是从实践开始的，现代建筑教育也将实践作为重要的教育环节。

中国的现代建筑和建筑教育在特殊的地缘政治环境中发展成长，同济大学建筑系在中国现代建筑史上占有重要的地位，被称之为“同济学派”。这个学派一贯注重跨学科的发展，坚持现代建筑的理性精神和现代建筑思想，创导博采众长，兼收并蓄的学术精神。作为中国建筑的同济学派，已经建立了丰厚的建筑理论体系和建筑思想，形成了独特的建筑教育体系，有着一批优秀的建筑教育家、建筑理论家和建筑师，培养了大量的人才，出现了一批优秀的建筑设计作品，都市建筑设计院的作品已经成为同济学派的重要组成部分。

作为依托同济大学建筑设计研究院和同济大学建筑与城市规划学院的学术研究、工程技术和人才培养优势的一所建筑设计院，都市建筑设计院的设计业务涵盖了建筑设计、室内设计、城市设计等，作品遍布中国大地。经过长期的锤炼和积累，创造了相当辉煌的业绩，许多作品获得国际、国内的各种奖项，一些作品已经成为重要的城市地标。设计的建筑类型涉及公共建筑、体育建筑、办公楼、大剧院、艺术馆、文化艺术中心、科研中心、教育建筑、商业建筑、住宅、博览建筑、观演建筑等。如同敏感的艺术家，老师们的许多作品表现出一种对这个世界的感悟和回应，正如法国哲学家莫里斯•梅洛－庞蒂所说：“建筑的任务是将这个世界如何触动我们的心弦展现出来。”

从《同济都市建筑十年》我们可以看到都市建筑设计院的建筑思想十分活跃，在目前的建设大潮中依然坚持实验性和先锋性，在作品中探索深层次的社会人文关怀，在设计中进行前瞻性的研究，全面吸收并学习多元的建筑理论和设计思潮，与国际建筑界和学术界的同行进行广泛的合作与交流，设计具有批判性意义的建筑，创造符合我们这个时代精神的作品。与一般的建筑教育院校不同，都市建筑设计院为同济大学建筑系的教师提供设计基地，并使设计从理想和图纸变成现实。作为大学的一部分，他们既从事建筑教学，培养研究生，也从事建筑设计，因而他们的视野更为宏观，更为国际化，也更具现代性，这些优势也体现在他们的作品中。

同济学派一直在建筑理论、建筑教育和建筑实践方面坚持不懈地探索中国新建筑的方向，执着地坚持现代建筑的理性精神和建筑教育思想，创导缜思的学风。也一直在探索中国实验性建筑的道路，从这十年的作品也可以看出建筑师们为社会发展服务，努力创造新的生活方式的特点，提倡跨学科和多学科的发展，关注生活，关注生态，关注城市更新，关注历史建筑和历史街区的保护等。他们的作品不追求宏大叙事，也不博眼球，不自娱自乐，而是崇尚真诚，反对英雄主义。他们在业主的需求、社会的需求、建筑师的专业水准和城市环境之间，实现学术研究、社会服务和人才培养的完美统一。在这本作品集中我们见到的是理性和严谨的设计，一种具有批判性意义的现代建筑。

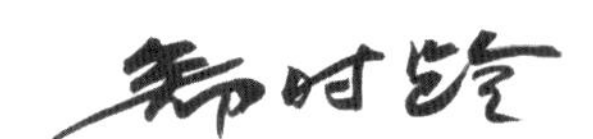

中国科学院院士

同济大学建筑设计研究院（集团）有限公司董事长

都市建筑设计院顾问

2015 年 11 月

PREFACE

It has been ten years ever since the launch of the Tongji Urban Architectural Design Institute , or TJUADI of TJAD (Tongji Architectural Design (Group) Co., Ltd.). Ten years are long enough for the maturity of an architectural design firm. As a design firm in the vicissitude of history, TJUADI dates back to the the Tongji Design Institute of Architecture and Civil Engineering founded in 1956. It was initiated to provide a practicing venue for teaching staff and students in the Department of Architecture. The world we are living in has been artificially reconstructed, full of creations with intentions. In this process architects' contribution is indispensable. By virtue of the nature of architects' practice, they undertake social obligations and should be responsible for the mistakes and malfunctions of buildings. They should also be responsible for the security of lives and assets, for the life quality of the masses and for overall well beings of the society. Architects in history were cultivated from practicing. In this sense, modern architectural education saw practicing as an integral component in training.

Modern architecture and modern architectural education in China grow in a special geopolitical environment. The Department of Architecture in Tongji University is called "Tongji School" with regard to its ranking in the modern architectural history of China. "Tongji School" maintains its commitment to the rationalism of modern architecture and pursues a doctrine of diversity. "Tongji School" has established a substantial theoretical system of architecture, has formed a distinct educational system, has built a teaching team including distinguished educators, theorists and practitioners, has cultivated numerous talented professionals and has produced a bunch of highly-evaluated design projects. TJUADI's contribution has become the integral part of "Tongji School."

As a design institute as well as research, education and training institute,relying on TJAD and CAUP, the services of TJUADI covers architectural design, interior design and urban design all over the country. After enduring accumulating and honing, TJUADI has made brilliant achievements with many projects receiving international and national awards. A couple of projects have become landmarks of their host cities. The buildings types include state or civic buildings, sports venues, offices, theaters, art museums, art centers, research headquarters, educational institutes, commercial buildings, residential buildings, exhibition halls and performance centers. Professors' works reflect their perception and reaction to the world as if they are acute artists.. As Maurice Merleau-Ponty remarks, "the task of architecture is to make visible how the world touches us."

From Selections from Design Works of Urban Architectural Design Institute we can observe the lively architectural thoughts of TJUADI and its leadership in the tide of construction in China. TJUADI explores embedded humane care in the projects, conducts advanced researches in design practice and comprehensively absorbs nutrients from diverse architectural theories. It seeks to broadly collaborate with international peers and produce designs with critical values and zeitgeist of our era. In contrast to other architectural colleges, TJUADI provides an outpost for professors' practices and facilitate the realization of blueprints and design visions. As a part of university, the professors are enabled to be involved in teaching, mentoring and practicing. Hence, their vision becomes more globalized and modernized and such a vision can be embodied in their design projects.

"Tongji School" has been constantly committed to the pursuit of a new path of Chinese architecture in architectural theory, education and practice. It sticks with the rationalism of modern architecture and architectural education and promulgates rigorous training methods. From the monograph we can behold architects' social obligations and investigations for creating a new life style. We can also sense their commitment to disciplinary development, human life, ecology, urban renovation and historical conservation. TJUADI's designers avoid being trapped in master narrative, spectacles or narcissism. They believe in honesty and object to heroism. They attempt to locate a balancing point between clients' expectation, social needs, professionalism and urban environment. They are looking for the integration of academic research, social service and education. This monograph represents a design principle of rationalism and rigor, and a modern architecture of critical meaning.

ZHENG Shiling
Academician of the Chinese Academy of Sciences
Chairman of the Board, Tongji Architectural Design (Group) Co., Ltd.
Consultant, Tongji Urban Architectural Design Institute
November 2015

前　言

2015年正值同济大学建筑与城市规划学院与同济大学建筑设计研究院（集团）有限公司在建筑设计领域的合作机构——都市建筑设计院成立十周年。十年来在各方的支持与关心下，都市建筑设计院在加强建筑教师创作实践平台建设、探索专业理论研究与社会应用实践的整合途径以及寻求建筑专业教学与设计行业发展的对接模式等方面取得了丰硕的成果，形成了一整套系统的运作机制与管理方式，与此同时，创作完成了一大批颇具社会影响且深得业主满意的优秀建筑作品。从创建落地到成长壮大，都市建筑设计院为更好地促进社会服务与人才培养、学术研究的有机统一，推动建筑学科的健康发展，走出了一条切合自身定位的发展之路。

同济大学建筑与城市规划学院前身是1952年全国高等院系调整时成立的同济大学建筑系，当时师资群体拥有英、美、德、法、奥、日等多国教育背景，阵容强大，思想活跃，学术氛围多元交融。其中不乏有冯纪忠、黄作燊、吴景祥、陈植、谭垣、黄家骅、哈雄文、庄秉权等中国早期建筑大家与优秀建筑师。学院自成立之初就坚持教学结合实践、服务社会的现代专业教育思想。1956年借鉴医学院办附属医院的模式，同济大学土木建筑设计院成立。设计院当时是“边教学边生产”，与教学联系十分密切。设计院成立前后的时期也正是同济建筑教师建筑创作实践最为辉煌的阶段之一，文远楼、同济工会俱乐部、花港茶室、同济大礼堂等一批经典作品，都突出体现了“中国的、当代的”原创精神，并成为探索中国现代建筑道路上的重要里程碑。

1978年在土木建筑设计院基础上成立了建筑设计研究院；之后2008年，同济大学建筑设计研究院（集团）有限公司揭牌。设计集团经过几代人的不懈努力，发展规模、服务能力及社会影响与日俱增，已走在了全国同类设计机构的前列。设计集团在各个时期与学院的合作和相互支持从未间断。进入新世纪，面对激烈的设计市场竞争，设计集团在加大品牌建设、实施“做大做强”战略、推行现代企业管理制度与设计质量标准中，一个适应学院教师进行工程实践与研究探索的专属平台也应运而生。作为学院与设计集团联合设置的合作机构，都市院既是学院教师在设计研究领域对外进行社会服务的窗口，同时它又是设计集团的一个分支机构，隶属学院和设计集团双重领导。都市院成立的基本宗旨是坚持产学研的协同发展，发挥学院与设计院双方各自的长处，优势互补、强强联合，共同提高学院和设计集团在建筑设计领域内的专业地位，更好地维护同济设计的社会声誉与责任担当。

都市院依托学院与设计集团的整体学术优势，拥有雄厚的设计实力与丰富的人才资源，设计研究范围涵盖了建筑设计、城市设计、历史建筑保护、室内设计、景观设计、环境艺术、特殊照明等多个领域。都市院的人员构成充分体现了以教师参与实践为主导的特点，主要由两部分组成。一部分为同济大学建筑与城市规划学院的在编教师，根据国家住建部、教育部的相关规定以及设计集团的相关办法，定期聘有70余位专业教师直接参与工程实践，他们中的半数以上具备教授或副教授等高级职称，绝大部分都拥有国家一级注册建筑师的执业资格，或者为各自研究领域的知名专家，他们是都市院最重要的建筑创作力量；另一部分人员为专职设计人员，经过多年专业磨练，他们正逐步成长为都市院的骨干力量，并已成为教师的得力助手与合作者。目前，都市院共设有：8个建筑设计室，两个城市设计室，两个环境设计室，历史建筑保护与再生、建筑集群与环境艺术等3个专业设计室，和高新技术设计研究所、建筑综合所。

近年来，都市院陆续承担了北京奥运会场馆、上海世博会场馆等一大批国家和地方建设的重大设计项目，并获得了各级政府的多种集体荣誉表彰与数百项国家级、省部级等各类优秀设计奖。其中，积极应对社会重大事件的设计任务，则充分体现了一个高校设计机构服务国家战略与社会民生的历史责任，也为同济建筑面向实践、面向社会的学科定位提供了最好的佐证。在2008年“5·12”汶川大地震发生后，都市院受上海市政府、同济大学以及学院和设计集团等的委托，担负了大量的灾区援建项目的建筑设计与设计组织工作。作为上海市的支援项目，都市院组织了全校建筑、规划、景观、结构、环境、交通、地质、水利、材料等相关专业的二十余位专家赴地震重灾区北川老县城现场调研，之后，项目团队又经过反复设计研讨与深化，最终完成了“北川国家地震遗址博物馆策划与整体方案设计”的一整套成果，为北川地震受灾区域与老县城地震遗址保护的总体定位、发展方向与决策操作提供了重要的指导依据。该项目分别获得了2009年度上海市、四川省和全国优秀城乡规划设计一等奖。与此同时，都市院还组织了二十余位学院教师领衔的设计团队赴上海市对口援建都江堰灾后建设现场，与设计集团其他部门通力合作完成了都江堰壹街区安居房灾后重建项目的设计工作。该工程的一批项目共获得了上海市优秀勘察设计工程奖、上海市建筑学会优秀创作奖、四川省“四花”优秀工程设计奖等十余项省部级表彰。都市院由此荣获了由上海市对口援建都江堰灾后重建指挥部授予的“优秀援建团体”称号。

建筑设计作品是设计机构综合管理水平与技术实践积累的真实反映，是其设计人员整体创作能力的集中体现。此次从十年以来都市院主创建筑师承担的逾千个项目中，选取近120项建成作品辑集成册，旨在检验过往的同时，加强与各界的专业互动与交流，以进一步促进都市院的发展，并提升其建筑创作实践的专业价值和学术影响力。这些作品，在项目类型上，包括了文化建筑、教育建筑、商业建筑、办公建筑、体育建筑、居住建筑、历史建筑保护和环境设计等不同类型，从中反映出都市院依托不同的学术团队研究基础而形成的建筑创作专长与特色；在设计规模上，既有几百平方米建筑面积的室内改建，也有上百公顷用地的城市设计，注重项目的创作价值与社会意义，使研究性设计机构的身份显现无疑；在创作阶段上，无论是项目方案设计、初步设计，还是全过程工程设计，均以项目实施中力求体现原创精神为根本追求与最终目标；在设计组织上，鉴于都市院以建筑为主的专业背景，积极发挥建筑创作的引领作用，加强与设计集团各技术专业的紧密合作已趋常态。此外，随着国际交流的日益深入，与境外顶尖设计机构与设计师在建筑项目创作上的合作也日渐增多，这在一定程度上也推动了同济建筑设计品牌的国际化。

百年积淀、十年一剑。都市建筑设计院依托同济大学深厚的历史文化底蕴，凭藉学院与设计集团的技术与政策支持，在建筑创作产学研协同发展道路上取得了长足的进步。注重实践、服务社会，是建筑学科发展的基本导向，是人才培养的根本要求，也是中国大学的重要职责与必然选择。站在新的起点上，都市院将继续以同济建筑学科奠基人冯纪忠先生“出人才，出作品，出思想”的办学理念作为自身的发展目标，以设计人才与设计思想的培育为至上追求，更多地面向社会、面向未来，加强实体化与规模化建设，拓展管理职能，进一步提升运行效率与整体实力，在坚持服务学科的同时，向符合专业发展规律的创新型设计机构方向转型，以更多、更好的作品贡献于社会。

《同济都市建筑十年》编委会

2015年10月

PROLOGUE

The Tongji Urban Architectural Design Institute (TJUADI) is celebrating its 10th anniversary! As a design institute collaboratively launched by the College of Architecture and Urban Planning (CAUP) and Tongji Architectural Design (Group) Co., Ltd. (TJAD), for 10 years TJUADI has made remarkable achievements in bridging the gap between teaching and practicing, integrating theory with praxis, and aligning professional training with design market. It has developed a kit of work flows and management approaches and at the same time completed a bunch of design projects with high user satisfaction and social recognition. TJUADI has found its own growth path by balancing social services, team building and academic researches. This path is leading to well-rounded development of the discipline of architecture.

The College of Architecture and Urban Planning (CAUP) dates back to the Department of Architecture and has a strenuous and energetic faculty with diverse education backgrounds from the United Kingdom, the United States, Germany, France, Austria and Japan. At the early stage of the college, its faculty includes master architects and educators such as Prof. FENG Jizhong, Prof. HUANG Zuoshen (Henry Huang), Prof. WU Jingxiang, Prof. CHEN Zhi, Prof. TAN Yuan, Prof. HUANG Jiahua, Prof. HA Xiongwen and Prof. ZHUANG Bingquan. Ever since its foundation, CAUP has been holding a doctrine that teaching should join practicing and serve the society. In 1956, the Tongji Design Institute of Architecture and Civil Engineering was established. Sticking to the "teaching-practicing" motto, the Tongji Design Institute marched to its heyday of production. A series of masterpieces, such as Wenyuan Hall, Tongji Union Club, Huagang Tea House, Tongji Grand Auditorium, became milestones in the history of Chinese modern architecture.

In 1978, the Tongji Design Institute of Architecture and Civil Engineering were reorganized as the Tongji Architectural Design Institute. In 2008, the Tongji Architectural Design (Group) Co., Ltd. (TJAD) were launched. Over a couple of generations, TJAD's growing volume, capacity and influences empowers it to be the vanguard among design institutes of such kind. In the new century, in response to escalating competition, TJAD adopted a "big-and-strong" strategy and kept applying modern enterprise management and design quality standards. A special platform for faculty members' practices and researches came into being. As an institute jointly established by CAUP and TJAD, TJUADI is a window for practicing the college's social service beyond its research tasks and as well as an outpost of TJAD. The principle of TJUADI is to unite practice, teaching and research and help CAUP and TJAD establish a social reputation and professional ranking which deserves its obligations.

TJUADI owns a distinguished workforce which is built around the academic vanguards of the college. The institute's scope of services covers architectural design, urban design, historical conservation, interior design, landscape architecture, environmental art and special lighting. The personnel of TJUADI is composed of two groups. One group are college faculty members. The majority of them are licensed architects or prestigious scholars, and half of them are tenured professors. The other group are designers who after years of training have gradually constituted the cadre of TJUADI and become qualified assistants and partners of the professors. Currently TJUADI comprises 8 architectural design studios, 2 urban design studios,

2 environmental design studios, a historical conservation studio, collective form studio and environmental art studio, a high-tech research institute and a comprehensive design studio.

Recently TJUADI has been commissioned to design a couple of sports venues of the Beijing 2008 Olympics and exhibition pavilions of the Shanghai World' s Fair and received variegated awards and prizes. As a collegiate design firm, TJUADI should take social obligations as a way to contribute to the society. After the 2008 Sichuan Earthquake, TJUADI was commissioned by Shanghai municipal government, Tongji University and the College of Architecture and Urban Planning to be in charge of many reconstruction projects. Led by TJUADI, a consulting team comprising architects, planners, landscape architects, civil engineers, traffic engineers, geologist, hydraulic engineers and material engineers were mobilized to conduct field survey in post-quake Beichuan County. After many rounds of vetting and discussion, a proposal for the Beichuan National Quake Museum and Memorial Park was submitted and this proposal provided comprehensive guidance and reference for the conservation of the after-quake ruins. Meanwhile, a group of design team including over 20 faculty members were organized to help the reconstruction of the city of Dujiangyan, another disaster area in the 2008 Earthquake. They completed the "Block One" resettlement project as one of the major reconstruction tasks. This project received many awards and accolades on national and provincial levels. Regarding TJUADI' s contribution in the after-quake reconstruction, it was awarded a title of "Distinguished Reconstruction Team" by the municipal government of Shanghai.

Built architectural projects honestly reflect the general administrative capacity and professional expertise of a design institute. The 120 built projects included in this monograph are selected from nearly a thousand pieces of work undertaken by TJUADI for the past 10 years. The compiling of this collection aims to help TJUADI communicate its design principles and promulgate the disciplinary value and academic influences of its practice. In terms of the range of services, the monograph cover diverse building types which reflect variegated research directions of the college. In terms of project sizes, the monograph contains projects ranging from interior design to large-scale urban design. In terms of design process management, TJUADI has established a working framework which both stresses the leading role of architects and enhances intimate collaboration with TJAD. Alongside deepening cooperation with top international design institutes and architects, "Tongji Architectural Design" as a label has been promoted to the global market.

Underlying TJUADI' s achievements in integrating research and practice is the thick history and culture of Tongji University and the technical and administrative assistance from CAUP and TJAD. "Serving the society" is the direction of the disciplinary development, the educational requirement and the foremost doctrine of all Chinese universities. On the threshold of a new chapter, TJUADI will keep implementing the motto of the college founder Prof. FENG Jizhong --- "education, creation, contemplation." It will continue to incubate human assets and upgrade its professional competitiveness. When taking hold of disciplinary contribution, TJUADI will follow a new path leading to an innovation-oriented design firm and will pay back the society with better services and works.

Editorial Board of the "*Selections from Design Works of Urban Architectural Design Institute*"

October 2015

目　录
CONTENS

北川地震纪念馆
Beichuan National Earthquake Memorial

项目名称：北川地震纪念馆
建筑地点：绵阳市北川羌族自治县
设计时间：2010 年
竣工时间：2014 年
设计阶段：方案设计、初步设计、施工图设计
建设单位：绵阳市唐家山堰塞湖治理暨北川老县城保护工作指挥部
基地面积：142 300m²
建筑面积：14 280m²
主要结构：钢筋混凝土框架结构
主要用途：展览、科研

Project：Beichuan National Earthquake Memorial
Construction Site：Beichuan Qiang Autonomous County, Mianyang
Design Period：2010
Completion：2014
Design Phase：Concept Design, Developments Design, Construction Documents Design
Client：Mianyang City Quake Lake Governance and the Old Town of Beichuan Protection Headquarters
Site Area：142,300m²
Floor Area：14,280m²
Main Structure：Reinforced Concrete Frame Structure
Main Application：Exhibition, Scientific Research

山东省美术馆
Shandong Art Museum

项目名称：山东省美术馆
建筑地点：山东济南
设计时间：2011 年
竣工时间：2013 年
设计阶段：方案设计、初步设计、施工图设计
建设单位：山东省文化厅
基地面积：20 700m^2
建筑面积：52 138m^2
主要结构：钢筋混凝土框架、剪力墙、钢桁架结构
主要用途：美术馆

Project：Shandong Art Museum
Construction Site：Jinan, Shandong
Design Period：2011
Completion：2013
Design Phase：Concept Design, Development Design, Construction Documents Design
Client：Shandong Provincial Department of Culture
Site Area：20,700m^2
Floor Area：52,138m^2
Main Structure：Reinforced Concrete Frame, Shear Wall Structure, Steel Truss
Main Application：Art Museum

洛阳博物馆
Luoyang Museum

项目名称：洛阳博物馆
建筑地点：河南洛阳
设计时间：2007 年
竣工时间：2009 年
设计阶段：方案设计、初步设计、施工图设计
建设单位：洛阳市文物局
基地面积：20hm²
建筑面积：43 654m²
主要结构：钢筋混凝土框架结构
主要用途：博物馆

Project：Luoyang Museum
Construction Site：Luoyang, Henan
Design Period：2007
Completion：2009
Design Phase：Schematic Design, Design Development, Construction Documents Design
Client：Luoyang Municipal Administration of Cultural Heritage
Site Area：20hm²
Floor Area：43,654m²
Main Structure：Reinforced Concrete Frame Structure
Main Application：Museum

南宁市昆仑关战役博物馆
The Kunlunguan Campaign Museum, Nanning

项目名称：南宁市昆仑关战役博物馆
建设地点：广西壮族自治区南宁市
设计时间：2006 年
竣工时间：2008 年
设计阶段：方案设计、扩初设计、施工图设计
建设单位：南宁市昆仑关战役保护管理委员会
基地面积：28 800m^2
建筑面积：4 154.5m^2
主要结构：钢筋混凝土框架结构
主要用途：展览

Project：The Kunlunguan Campaign Museum, Nanning
Construction Site：Guangxi, Nanning
Design Period：2006
Completion：2008
Design Phase：Concept Design, Development Design, Construction Documents Design
Client：Nanning City Kunlunguan Campaign Protection Management Committee
Site Area：28,800m^2
Floor Area：4,154.5m^2
Main Structure：Reinforced Concrete Frame Structure
Main Application：Exhibition

阖闾城遗址博物馆
Helu City Historic Site Museum

项目名称：阖闾城遗址博物馆	Project：Helu City Historic Site Museum
建筑地点：江苏无锡	Construction Site：Wuxi, Jiangsu
设计时间：2009 年	Design Period：2009
竣工时间：2014 年	Completion：2014
设计阶段：方案设计、初步设计、施工图设计	Design Phase：Schematic Design, Design Development, Construction Documents Design
建设单位：无锡吴都阖闾古城发展有限公司	Client：Wuxi Wudu Helu City Development Co., Ltd.
基地面积：46 990m^2	Site Area：46,990m^2
建筑面积：26 526m^2	Floor Area：26,526m^2
主要结构：钢筋混凝土框架结构	Main Structure：Reinforced Concrete Frame Structure
主要用途：博物馆	Main Application：Museum

项目名称：上海会馆史陈列馆	Project：Shanghai Guildhall History Museum
建设地点：上海市	Construction Site：Shanghai
设计时间：2009 年	Design Period：2009
竣工时间：2010 年	Completion：2010
设计阶段：方案设计、初步设计、施工图设计	Design Phase：Concept Design, Development Design, Construction Documents Design
建设单位：上海黄浦区文化局	Client：Huangpu District Cultural Bureau, Shanghai
基地面积：3 897m^2	Site Area：3,897m^2
建筑面积：1 956m^2	Floor Area：1,956m^2
主要结构：钢筋混凝土框架结构、钢结构	Main Structure：Reinforced Concrete Frame Structure, Steel Frame Structure
主要用途：展览馆	Main Appliction：Exhibition Hall

上海会馆史陈列馆
Shanghai Guildhall History Museum

ZHU QIZHAN ART MUSEUM

项目名称：朱屺瞻艺术馆
建设地点：上海市
设计时间：2004 年
竣工时间：2005 年
设计阶段：方案设计、初步设计、施工图设计
建设单位：朱屺瞻艺术馆
基地面积：672.8m^2
建筑面积：2 083.2m^2
主要结构：混凝土框架、钢结构
主要用途：展览馆

Project：Zhu Qizhan Art Museum
Construction Site：Shanghai
Design Period：2004
Completion：2005
Design Phase：Concept Design, Development Design, Construction Documents Design
Client：Zhu Qizhan Art Museum
Site Area：672.8m^2
Floor Area：2,083.2m^2
Main Structure：Reinforced Concrete Frame Structure, Steel Frame Structure
Main Appliction：Exhibition Hall

朱屺瞻艺术馆
Zhu Qizhan Art Museum

项目名称：范曾艺术馆	Project：Fanzeng Art Museum
建设地点：南通市南通大学	Construction Site：Nantong University, Nantong
设计时间：2010 年	Design Period：2010
竣工时间：2014 年	Completion：2014
设计阶段：方案设计、初步设计、施工图设计	Design Phase：Concept Design, Development Design, Construction Documents Design
建设单位：南通大学	Client：Nantong University
基地面积：20 529m^2	Site Area：20,529m^2
建筑面积：7 028m^2	Floor Area：7,028m^2
主要结构：钢筋混凝土框架结构	Main Structure：Reinforced Concrete Frame Structure
主要用途：展览馆	Main Appliction：Exhibition Hall

范曾艺术馆
Fanzeng Art Museum

项目名称：上海交响乐团迁建工程
建筑地点：上海
设计时间：2008 年
竣工时间：2013 年
设计阶段：施工图设计
合作单位：矶崎新设计工作室
建设单位：上海交响乐团
基地面积：16 138m^2
建筑面积：19 950m^2
主要结构：钢筋混凝土框架结构
主要用途：文化

Project：New Concert Hall of Shanghai Symphony Orchestra
Construction Site：Shanghai
Design Period：2008
Completion：2013
Design Phase：Construction Documents Design
Partnership：Arata Isozaki & Associates
Client：Shanghai Symphony Orchestra
Site Area：16,138m^2
Floor Area：19,950m^2
Main Structure：Reinforced Concrete Frame Structure
Main Application：Culture

上海交响乐团迁建工程
New Concert Hall of Shanghai Symphony Orchestra

项目名称：东湖国际会议中心
建设地点：武汉市武昌区
设计时间：2008 年
竣工时间：2012 年
设计阶段：方案设计、初步设计、施工图设计
建设单位：湖北武汉东湖会议中心有限责任公司
基地面积：174 968m²
建筑面积：61 436m²
主要结构：钢筋混凝土框架结构、钢结构
主要用途：酒店、会议、餐饮

Project：Wuhan East Lake Hotel
Construction Site：Wuchang District, Wuhan
Design Period：2008
Completion：2012
Design Phase：Concept Design, Development Design, Construction Documents Design
Client：Wuhan East Lake Hubei Conference Center Co., Ltd.
Site Area：174,968m²
Floor Area：61,436m²
Main Structure：Reinforced Concrete Frame Structure, Steel Frame Structure
Main Appliction：Hotel, Conference, Catering

东湖国际会议中心
Wuhan East Lake Hotel

都江堰市文化馆

House of Culture, Dujiangyan

项目名称：都江堰市文化馆
建筑地点：四川省都江堰市
设计时间：2008 年
竣工时间：2010 年
设计阶段：方案设计、初步设计、施工图设计
建设单位：都江堰新城建设投资有限公司
基地面积：4 300m^2
建筑面积：5 500m^2
主要结构：钢筋混凝土框架结构
主要用途：文化

Project：House of Culture, Dujiangyan
Construction Site：Dujiangyan, Sichuan
Design Period：2008
Completion：2010
Design Phase：Concept Design, Development Design, Construction Documents Design
Client：Dujiangyan New City Construction Investment Co., Ltd.
Site Area：4,300m^2
Floor Area：5,500m^2
Main Structure：Reinforced Concrete Frame Structure
Main Application：Culture

都江堰市工人活动中心
Culture Activity Centre for Workers, Dujiangyan

项目名称：都江堰市工人活动中心
建设地点：四川省都江堰市
设计时间：2008 年
竣工时间：2010 年
设计阶段：方案设计、初步设计、施工图设计
建设单位：都江堰新城建设投资有限公司
基地面积：8 551m^2
建筑面积：13 806m^2
主要结构：钢筋混凝土框架
主要用途：文化

Project：Culture Activity Centre for Workers, Dujiangyan
Construction Site：Dujiangyan, Sichuan
Design Period：2008
Completion：2010
Design Phase：Concept Design, Development Design, Construction Documents Design
Client：Dujiangyan New City Construction Investment Co., Ltd.
Site Area：8,551m^2
Floor Area：13,806m^2
Main Structure：Reinforced Concrete Frame Structure
Main Application：Culture

都江堰市妇女儿童活动中心
Women's Child's Activity Centre, Dujiangyan

项目名称：都江堰市妇女儿童活动中心
建筑地点：四川省都江堰市
设计时间：2008 年
竣工时间：2010 年
设计阶段：方案设计、初步设计、施工图设计
建设单位：都江堰新城建设投资有限公司
基地面积：6 500m²
建筑面积：8 314m²
主要结构：钢筋混凝土框架结构
主要用途：文化

Project：Women's Child's Activity Centre, Dujiangyan
Construction Site：Dujiangyan, Sichuan
Design Period：2008
Completion：2010
Design Phase：Concept Design, Development Design, Construction Documents Design
Client：Dujiangyan New City Construction Investment Co., Ltd.
Site Area：6,500m²
Floor Area：8,314m²
Main Structure：Reinforced Concrete Frame Structure
Main Application：Culture

上海—都江堰产业转移促进中心
Shanghai-Dujiangyan Industry Transfer Promotion Center
都江堰市青少年宫
DUJIANGYAN YOUTH PALACE
青少年活动中心
YOUTH CENTER

杭州市萧山区图书馆文化馆
The Library and Culture Center of Xiaoshan District, Hangzhou

项目名称：杭州市萧山区图书馆文化馆
建筑地点：杭州市萧山区
设计时间：2005 年
竣工时间：2009 年
设计阶段：方案设计、初步设计、施工图设计
建设单位：萧山区政府
基地面积：17 945m^2
建筑面积：32 011m^2
主要结构：钢筋混凝土框架结构
主要用途：文化

Project：The Library and Culture Center of Xiaoshan District, Hangzhou
Construction Site：Xiaoshan, District, Hangzhou
Design Period：2005
Completion：2009
Design Phase：Concept Design, Development Design, Construction Documents Design
Client：The government of Xiaoshan District
Site Area：17,945m^2
Floor Area：32,011m^2
Main Structure：Reinforced Concrete Frame Structure
Main Application：Culture

、法治，爱国、敬业、诚信

淄博市张店区文化艺术中心
Culture and Art Center of Zhangdian District, Zibo

项目名称：淄博市张店区文化艺术中心
建筑地点：淄博市张店区
设计时间：2005 年
竣工时间：2006 年
设计阶段：方案设计
建设单位：淄博市张店区文化局
基地面积：5 775m²
建筑面积：10 612m²
主要结构：钢筋混凝土框架结构
主要用途：文化

Project：Culture and Art Center of Zhangdian District, Zibo
Construction Site：Zhangdian District, Zibo
Design Period：2005
Completion：2006
Design Phase：Concept Design
Client：Zhangdian District Bureau of Culture, Zibo
Site Area：5,775m²
Floor Area：10,612m²
Main Structure：Reinforced Concrete Frame Structure
Main Application：Culture

项目名称：晋中市城市规划展示馆
建设地点：山西省晋中市
设计时间：2012 年
竣工时间：2014 年
设计阶段：方案设计、初步设计、施工图设计
建设单位：晋中市规划局
基地面积：25 100m^2
建筑面积：17 451m^2
主要结构：钢筋混凝土框架结构
主要用途：展示馆

Project：Urban Planning Exhibition Hall in Jinzhong
Construction Site：Jinzhong, Shanxi Province
Design Period：2012
Completion：2014
Design Phase：Concept Design, Development Design, Construction Documents Design
Client：Jinzhong Planning Bureau
Site Area：25,100m^2
Floor Area：17,451m^2
Main Structure：Reinforced Concrete Frame Structure
Main Appliction：Exhibition Hall

晋中市城市规划展示馆
Urban Planning Exhibition Hall in Jinzhong

项目名称：无锡市惠山区建设与发展展示中心
建设地点：无锡市惠山区
设计时间：2006 年
竣工时间：2008 年
设计阶段：方案设计、初步设计、施工图设计
建设单位：无锡惠新资产经营管理有限公司
基地面积：30 941m^2
建筑面积：15 055m^2
主要结构：钢筋混凝土框架结构
主要用途：展览馆

Project：The Constrution & Development Exhibition Hall of the Huishan District, Wuxi
Construction Site：Huishan District, Wuxi
Design Period：2006
Completion：2008
Design Phase：Concept Design, Development Design, Construction Documents Design
Client：Wuxi Huixin Asset Management Co., Ltd.
Site Area：30 941m^2
Floor Area：15 055m^2
Main Structure：Reinforced Concrete Frame Structure
Main Appliction：Exhibition Hall

无锡市惠山区建设与发展展示中心

The Constrution & Development Exhibition Hall of the Huishan District, Wuxi

烟台市经济技术开发区城市规划中心
Urban Planning Centre of Yantai Economic & Technological Development Area

项目名称：烟台市经济技术开发区城市规划中心
建设地点：山东省烟台市
设计时间：2012 年
竣工时间：2014 年
设计阶段：方案设计、初步设计、施工图设计
建设单位：烟台经济技术开发区规划局
基地面积：34 410.6m^2
建筑面积：19 892m^2
主要结构：钢筋混凝土框架结构
主要用途：展览 办公

Project：Urban Planning Centre of Yantai Economic & Technological Development Area
Construction Site：Yantai, Shandong Province
Design Period：2012
Completion：2014
Design Phase：Concept Design, Development Design, Construction Documents Design
Client：Yantai Economic & Technological Development Planning Bureau
Site Area：34,410.6m^2
Floor Area：19,892m^2
Main Structure：Reinforced Concrete Frame Structure
Main Appliction：Exhibition Hall and Office

中国城市化史馆·清河文展中心
The Museum of Chinese Urbamization & Qinghe Culture and Exhibition Center

项目名称：中国城市化史馆·清河文展中心
建筑地点：江苏省淮安市
设计时间：2009 年
竣工时间：2011 年
设计阶段：方案设计、初步设计、施工图设计
建设单位：淮安清河新区投资发展有限公司
基地面积：24 318m^2
建筑面积：35 348m^2
主要结构：钢筋混凝土框架结构，局部钢结构
主要用途：展示

Project：The Museum of Chinese Urbamization & Qinghe Culture and Exhibition Center
Construction Site：Huai'an, Jiangsu Province
Design Period：2009
Completion：2011
Design Phase：Concept Design, Development Design, Construction Documents Design
Client：Huai'an Qinghe New District Investment & Development Co., Ltd.
Site Area：24,318m^2
Floor Area：35,348m^2
Main Structure：Reinforced Concrete Frame Structure, Steel Frame Structure
Main Application：Exhibition

扬中市中华河豚岛观光塔
Tower of Globefish Island of China, Yangzhong

项目名称：扬中市中华河豚岛观光塔
建筑地点：江苏省扬中市
设计时间：2012 年
竣工时间：2013 年
建设单位：江苏天禾旅游发展有限公司
设计阶段：方案设计、施工图设计
基地面积：13 000m^2
建筑面积：2 000m^2
主要结构：钢结构
主要用途：观光、展示

Project：Tower of Globefish Island of China, Yangzhong
Construction Site：Yangzhong, Jiangsu
Design Period：2012
Completion：2013
Client：Jiangsu Tianhe Tourism Development Co., Ltd.
Design Phase：Concept Design, Construction Documents Design
Site Area：13,000m^2
Floor Area：2,000m^2
Main Structure：Steel Structure
Main Application：Sightseeing, Exhibition

浙江大学紫金港校区中心岛建筑群

The Building Complex on Central Island in Zijingang Campus of Zhejiang University

项目名称：浙江大学紫金港校区中心岛建筑群
建设地点：浙江大学紫金港校区
设计时间：2003 年
竣工时间：2005 年
设计阶段：方案设计、初步设计、施工图设计
建设单位：浙江大学
基地面积：46 115m^2
建筑面积：40 000m^2
主要结构：钢筋混凝土框架结构、钢结构
主要用途：会议、办公

Project：The Building Complex on Central Island in Zijingang Campus of Zhejiang University
Construction Site：Zijingang Campus, Zhejiang University
Design Period：2003
Completion：2005
Design Phase：Concept Design, Development Design, Construction Documents Design
Client：Zhejiang University
Site Area：46,115m^2
Floor Area：40,000m^2
Main Structure：Reinforced Concrete Frame Structure, Steel Frame Structure
Main Application：Meeting Office

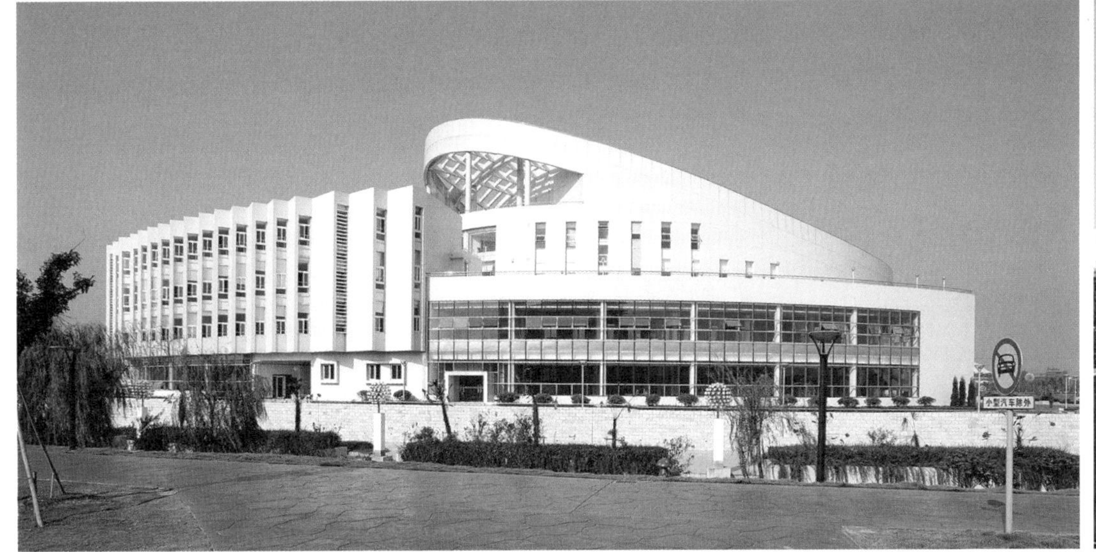

项目名称：上海音乐学院教学楼
建筑地点：上海
设计时间：2005 年
竣工时间：2007 年
设计阶段：方案设计、初步设计、施工图设计
建设单位：上海市静安区教育局
基地面积：15 200m^2
建筑面积：24 700m^2
主要结构：钢筋混凝土框架结构
主要用途：教育

Project：The Educational Building of Conservatory of Music, Shanghai
Construction Site：Shanghai
Design Period：2005
Completion：2007
Design Phase：Concept Design, Development Design, Construction Documents Design
Client：Bureau of Fducation of Jing'an District, Shanghai
Site Area：15,200m^2
Floor Area：24,700m^2
Main Structure：Reinforced Concrete Frame Structure
Main Application：Education

上海音乐学院教学楼
The Educational Building of Conservatory of Music, Shanghai

项目名称：上海音乐学院排演中心
建筑地点：上海
设计时间：2005 年
竣工时间：2007 年
设计阶段：方案设计、初步设计、施工图设计
建设单位：上海市静安区教育局
基地面积：1 850m²
建筑面积：4 997m²
主要结构：钢筋混凝土框架结构
主要用途：教育

Project：The Rehearsal Hall of Conservatory of Music, Shanghai
Construction Site：Shanghai
Design Period：2005
Completion：2007
Design Phase：Concept Design, Development Design, Construction Documents Design
Client：Bureau of education of Jing'an District, Shanghai
Site Area：1,850m²
Floor Area：4,997m²
Main Structure：Reinforced Concrete Frame Structure
Main Application：Education

上海音乐学院排演中心
The Rehearsal Hall of Conservatory of Music, Shanghai

项目名称：上海音乐学院实验学校
建设地点：上海新江湾城
设计时间：2007 年
竣工时间：2009 年
建设单位：上海市城市建设投资开发总公司
基地面积：33 046m^2
建筑面积：25 426m^2
主要结构：钢筋混凝土框架结构、钢结构
主要用途：教育

Project：Experimental School of Shanghai Conservatory of Music
Construction Site：New Jiangwan City, Shanghai
Design Period：2007
Completion：2009
Client：Shanghai Urban Construction (Groop) Corporation
Site Area：33,046m^2
Floor Area：25,426m^2
Main Structure：Reinforced Concrete Frame Structure, Steel Frame Structure
Main Appliction：Education

上海音乐学院实验学校
Experimental School of Shanghai Conservatory of Music

同济大学土木工程学院大楼
The New Building of School of Civil Engineering, Tongji University

项目名称：同济大学土木工程学院大楼
建筑地点：上海市杨浦区
设计时间：2005 年
竣工时间：2006 年
设计阶段：方案设计、初步设计、施工图设计
建设单位：同济大学基建工程部
基地面积：10 006m^2
建筑面积：14 920m^2
主要结构：钢框架结构
主要用途：教育

Project：The New Building of School of Civil Engineering, Tongji University
Construction Site：Yangpu District, Shanghai
Design Period：2005
Completion：2006
Design Phase：Concept Design, Developments Design, Construction Documents Design
Client：Department of Infrastructure Engineering, Tongji University
Site Area：10,006m^2
Floor Area：14,920m^2
Main Structure：Steel Frame Structure
Main Application：Education

同济大学嘉定校区留学生宿舍及专家公寓

Foreign Students' Dormitory and Expert Apartments in Jiading Campus of Tongji University

项目名称：同济大学嘉定校区留学生宿舍及专家公寓
建设地点：上海市嘉定区
设计时间：2011 年
竣工时间：2014 年
设计阶段：方案设计、初步设计、施工图设计
建设单位：同济大学
基地面积：22 143m^2
建筑面积：35 870m^2
主要结构：钢筋混凝土框架结构
主要用途：教育

Project：Foreign Students'Dormitory and Expert Apartments in Jiading Campus of Tongji University
Construction Site：Jiading, Shanghai
Design Period：2011
Completion：2014
Design Phase：Concept Design, Development Design, Construction Documents Design
Client：Tongji University
Site Area：22,143m^2
Floor Area：36,870m^2
Main Structure：Reinforced Concrete Frame Structure
Main Application：Education

重庆交通大学双福校区
Shuangfu Campus of Chongqing Jiaotong University

项目名称：重庆交通大学双福校区
建筑地点：重庆
设计时间：2009 年
竣工时间：2011 年
设计阶段：方案设计
建设单位：重庆交通大学
基地面积：20hm^2
建筑面积：176 352m^2
主要结构：钢筋混凝土框架结构
主要用途：教育

Project：Shuangfu Campus of Chongqing Jiaotong University
Construction Site：Chongqing
Design Period：2009
Completion：2011
Design Phase： Detailed Planning, Concept Design
Client：Chongqing Jiaotong University
Site Area：20hm^2
Floor Area：176,352m^2
Main Structure：Reinforced Concrete Frame Structure
Main Application：Education

语音计算机楼
Language and Computer Labs Building

B01

中国石油大学青岛校区
Qingdao Campus of China University of Petroleum

项目名称：中国石油大学青岛校区	Project：Qingdao Campus of China University of Petroleum
建筑地点：山东省青岛市	Construction Site：Qingdao Shandong
设计时间：2005 年	Design Period：2005
竣工时间：2007 年	Completion：2007
设计阶段：修建性详细规划、方案设计	Design Phase：Detailed Planning, Concept Design
建设单位：石油大学（华东）	Client：Petroleum University (Hua Dong)
基地面积：36hm^2	Site Area：36hm^2
建筑面积：290 000m^2	Floor Area：290,000m^2
主要结构：钢筋混凝土框架结构	Main Structure：Reinforced Concrete Frame Structure
主要用途：教育	Main Application：Education

日照职业技术学院新校区

New Campus of Rizhao Polytechnic

项目名称：日照职业技术学院新校区
建筑地点：山东省日照市
设计时间：2005 年
竣工时间：2007 年
设计阶段：修建性详细规划、方案设计
建设单位：日照职业技术学院
基地面积：$15hm^2$
建筑面积：$118\,000m^2$
主要结构：钢筋混凝土框架结构
主要用途：教育

Project：New Campus of Rizhao Polytechnic
Construction Site：Rizhao Shandong
Design Period：2003
Completion：2007
Design Phase：Detailed Planning，Concept Design
Client：Rizhao Polytechnic
Site Area：$15hm^2$
Floor Area：$118,000m^2$
Main Structure：Reinforced Concrete Frame Structure
Main Application：Education

项目名称：上海宣传党校迁建工程
建筑地点：上海
设计时间：2008 年
竣工时间：2014 年
设计阶段：方案设计、初步设计、施工图设计
建设单位：中共上海市委宣传部
基地面积：63 910m^2
建筑面积：11 613m^2
主要结构：钢筋混凝土框架结构
主要用途：教育

Project：The Migratory Construction of the Party School of Propaganda Department, Shanghai
Construction Site：Shanghai
Design Period：2008
Completion：2014
Design Phase：Concept Design, Development Design, Construction Documents Design
Client：Propaganda Department of CCP, Shanghai
Site Area：63,910m^2
Floor Area：11,613m^2
Main Structure：Reinforced Concrete Frame Structure
Main Application：Education

上海宣传党校迁建工程
The Migratory Construction of the Party School of Propaganda Department, Shanghai

项目名称：格致中学二期	Project：Gezhi Middle School（the 2nd Construction Stage）
建设地点：上海	Construction Site：Shanghai
设计时间：2003 年	Design Period：2003
竣工时间：2005 年	Completion：2005
设计阶段：	Design Phase：
建设单位：黄浦区教育局	Client：Bureau of Educationof Huangpu District, Shanghai
基地面积：13 125m^2	Site Area：13,125m^2
建筑面积：17 417m^2	Floor Area：17,417m^2
主要结构：钢筋混凝土框架结构	Main Structure：Reinforced Concrete Frame Structure
主要用途：教育	Main Appliction：Education

格致中学二期
Gezhi Middle School（the 2nd Construction Stage）

上海市格致中学

项目名称：上海市市西中学改扩建工程	Project：The Reconstruction and Extension of Shixi Middle School, Shanghai
建筑地点：上海	Construction Site：Shanghai
设计时间：2007 年	Design Period：2007
竣工时间：2012 年	Completion：2012
设计阶段：方案设计、初步设计、施工图设计	Design Phase：Concept Design, Development Design, Construction Documents Design
建设单位：上海市静安区教育局	Client：Bureau of Education of Jing'an District, Shanghai
基地面积：6 200m²	Site Area：6,200m²
建筑面积：20 234m²	Floor Area：20,234m²
主要结构：钢筋混凝土框架结构	Main Structure：Reinforced Concrete Frame Structure
主要用途：教育	Main Application：Education

上海市市西中学改扩建工程

The Reconstruction and Extension of Shixi Middle School, Shanghai

青岛第二初级实验中学
The Second Experiment Middle School, Qingdao

项目名称：青岛第二初级实验中学
建筑地点：山东省青岛市
设计时间：2007—2008 年
竣工时间：2010 年
设计阶段：方案设计、初步设计、施工图设计
合作单位：青岛志海工程设计咨询公司
建设单位：山东鲁能置业有限公司
基地面积：34 450m²
建筑面积：24 892m²
主要结构：钢筋混凝土框架结构
主要用途：教育

Project：The Second Experiment Middle School, Qingdao
Construction Site：Qingdao Shandong
Design Period：2007-2008
Completion：2010
Design Phase：Concept Design, Developments Design, Construction Documents Design
Partnership：Qingdao Zhihai Engineering Design Consulting Co. Ltd.
Client：Shandong Luneng Real Estate Co., Ltd.
Site Area：34,450m²
Floor Area：24,892m²
Main Structure：Reinforced Concrete Frame Structure
Main Application：Education

项目名称：淄博市张店区第八中学
建筑地点：淄博市张店区
设计时间：2009 年
竣工时间：2011 年
设计阶段：修建性详细规划、方案设计
建设单位：淄博市张店区教育局
基地面积：83 405m^2
建筑面积：30 138m^2
主要结构：钢筋混凝土框架结构
主要用途：教育

Project：The Eighth Middle School of Zhangdian District, Zibo
Construction Site：Zhangdian District, Zibo
Design Period：2009
Completion：2011
Design Phase：Detailed Planning, Concept Design
Client：Zhangdian District Bureau of Education, Zibo
Site Area：83,405m^2
Floor Area：30,138m^2
Main Structure：Reinforced Concrete Frame Structure
Main Application：Education

淄博市张店区第八中学
The Eighth Middle School of Zhangdian District, Zibo

淄博市周村区第三中学
The Third Middle School of Zhoucun District, Zibo

项目名称：淄博市周村区第三中学
建筑地点：淄博市周村区
设计时间：2008 年
竣工时间：2010 年
设计阶段：修建性详细规划、方案设计
建设单位：淄博市周村区教育局
基地面积：66 700m^2
建筑面积：40 500m^2
主要结构：钢筋混凝土框架结构
主要用途：教育

Project：The Third Middle School of Zhoucun District, Zibo
Construction Site：Zhoucun District, Zibo
Design Period：2008
Completion：2010
Design Phase：Detailed Planning, Concept Design
Client：Zhoucun District Bureau of Education, Zibo
Site Area：66,700m^2
Floor Area：40,500m^2
Main Structure：Reinforced Concrete Frame Structure
Main Application：Education

淄博市般阳中学
Panyang Middle School, Zibo

项目名称：淄博市般阳中学
建筑地点：淄博市淄川区
设计时间：2011 年
竣工时间：2013 年
设计阶段：修建性详细规划、方案设计
建设单位：淄博市教育局
基地面积：174 600m^2
建筑面积：92 396m^2
主要结构：钢筋混凝土框架结构
主要用途：教育

Project：Panyang Middle School, Zibo Construction
Site：Zichuan District, Zibo
Design Period：2011
Completion：2013
Design Phase：Detailed Planning, Concept Design
Client：Zibo Municipal Bureau of Education
Site Area：174,600m^2
Floor Area：92,396m^2
Main Structure：Reinforced Concrete Frame Structure
Main Application：Education

淄博市淄博中学
Zibo Middle School, Zibo

项目名称：淄博市淄博中学
建筑地点：淄博市张店区
设计时间：2010 年
竣工时间：2012 年
设计阶段：修建性详细规划、方案设计
建设单位：淄博市教育局
基地面积：224 000m^2
建筑面积：98 500m^2
主要结构：钢筋混凝土框架结构
主要用途：教育

Project：Zibo Middle School, Zibo
Construction Site：Zhangdian District , Zibo
Design Period：2010
Completion：2012
Design Phase：Detailed Planning, Concept Design
Client：Zibo Municipal Bureau of Education
Site Area：224,000m^2
Floor Area：98,500m^2
Main Structure：Reinforced Concrete Frame Structure
Main Application：Education

杭州江南实验学校
Jiangnan Experimental School, Hangzhou

项目名称：杭州江南实验学校
建筑地点：浙江省杭州市
设计时间：2002 年
竣工时间：2008 年
设计阶段：方案设计、初步设计、施工图设计
建设单位：杭州市滨江区教育局
基地面积：82 850m^2
建筑面积：70 680m^2
主要结构：钢筋混凝土框架
主要用途：教育

Project：Jiangnan Experimental School, Hangzhou
Construction Site：Hangzhou, Zhejiang
Design Period：2002
Completion：2008
Design Phase：Concept Design, Development Design, Construction Documents Design
Client：Binjiang district Bureau of education, Hangzhou
Site Area：82,850m^2
Floor Area：70,680m^2
Main Structure：Reinforced Concrete Frame Structure
Main Application：Education

黄浦区第一中心小学
The First Central Primary School of Huangpu District

项目名称：黄浦区第一中心小学
建设地点：上海市
设计时间：2006 年
竣工时间：2009 年
设计阶段：方案设计、初步设计、施工图设计
建设单位：黄浦区教育局
基地面积：5 325m²
建筑面积：10 456m²
主要结构：钢筋混凝土框架结构、钢结构
主要用途：教育

Project：The First Central Primary School of Huangpu District
Construction Site：Huangpu District, Shanghai
Design Period：2006
Completion：2009
Design Phase：Concept Design, Development Design, Construction Documents Design
Client：Huangpu District Bureau of Education, Shanghai
Site Area：5,325m²
Floor Area：10,456m²
Main Structure：Reinforced Concrete Frame Structure, Steel Frame Structure
Main Appliction：Education

项目名称：淄博市周村区正阳小学
建筑地点：淄博市周村区
设计时间：2013 年
竣工时间：2015 年
设计阶段：修建性详细规划、方案设计
建设单位：淄博市周村区教育局
基地面积：30 503m^2
建筑面积：20 806m^2
主要结构：钢筋混凝土框架结构
主要用途：教育

Project：Zhengyang Primary School of Zhoucun District, Zibo
Construction Site：Zhoucun District, Zibo
Design Period：2013
Completion：2015
Design Phase：Detailed Planning, Concept Design
Client：Zhoucun District Bureau of Education, Zibo
Site Area：30,503m^2
Floor Area：20,806m^2
Main Structure：Reinforced Concrete Frame Structure
Main Application：Education

淄博市周村区正阳小学
Zhengyang Primary School of Zhoucun District, Zibo

淄博市张店区祥瑞园小学
Xiangruiyuan Primary School of Zhangdian District, Zibo

项目名称：淄博市张店区祥瑞园小学
建筑地点：淄博市张店区
设计时间：2008 年
竣工时间：2009 年
设计阶段：修建性详细规划、方案设计
建设单位：淄博市张店区教育局
基地面积：40 000m²
建筑面积：16 990m²
主要结构：钢筋混凝土框架结构
主要用途：教育

Project：Xiangruiyuan Primary School of Zhangdian District, Zibo
Construction Site：Zhangdian District, Zibo
Design Period：2008
Completion：2009
Design Phase：Detailed Planning, Concept Design
Client：Zhangdian District Bureau of Education, Zibo
Site Area：40,000m²
Floor Area：16,990m²
Main Structure：Reinforced Concrete Frame Structure
Main Application：Education

杭州市民中心
Hangzhou Civic Center

项目名称：杭州市民中心
建设地点：浙江省杭州市
设计时间：2002—2004 年
竣工时间：2012 年
设计阶段：方案设计、初步设计、施工图设计
建设单位：浙江建设工程集团有限公司
基地面积：176 189m^2
建筑面积：499 977m^2
主要结构：钢结构
主要用途：办公综合体

Project：Hangzhou Civic Center
Construction Site：Hangzhou, Zhejiang
Design Period：2002-2004
Completion：2012
Design Phase：Concept Design, Development Design, Construction Documents Design
Client：Zhejiang Construction Engineering Group Co., Ltd
Site Area：176,189m^2
Floor Area：499,977m^2
Main structure：Steel Structure
Main Application：Complex of Office

合肥市包河政务中心
Baohe District Administration Center, Hefei

项目名称：合肥市包河政务中心
建筑地点：安徽合肥
设计时间：2003—2005 年
竣工时间：2006 年
设计阶段：方案设计、初步设计、施工图设计
合作单位：合肥新站综合开发试验区规划建筑设计研究院
建设单位：包河区行政办公中心建设指挥部办公室
基地面积：35 500m²
建筑面积：44 780m²
主要结构：钢筋混凝土框架结构
主要用途：办公

Project：Baohe District Administration Center, Hefei
Construction Site：Hefei, Anhui Province
Design Period：2003-2005
Completion：2006
Design Phase：Concept Design, Development Design, Construction Documents Design
Partnership：Hefei Xinzhan Institute of Architectural Design & Planning
Client：Construction Office of Baohe Administration Center
Site Area：35,500m²
Floor Area：44,780m²
Main Structure：Reinforced Concrete Frame Structure
Main Application：Office

武汉市汉阳区机关大楼及行政办公中心
Administration Center of Hanyang District, Wuhan

项目名称：武汉市汉阳区机关大楼及行政办公中心
建筑地点：武汉市汉阳区
设计时间：2006 年
竣工时间：2008 年
设计阶段：方案设计、初步设计
建设单位：武汉市汉阳区机关事务管理局
基地面积：25 600m^2
建筑面积：37 051m^2
主要结构：钢筋混凝土框架结构
主要用途：办公

Project：Administration Center of Hanyang District, Wuhan
Construction Site：Hanyang District, Wuhan
Design Period：2006
Completion：2008
Design Phase：Concept Design , Developments Design
Client：Government Offices Administration of Hanyang District, Wuhan
Site Area：25,600m^2
Floor Area：37,051m^2
Main Structure：Reinforced Concrete Frame Structure
Main Application：Office

山东省博兴县行政大楼
Administration Building of Boxing County, Shandong

项目名称：山东省博兴县行政大楼
建筑地点：山东省博兴县
设计时间：2005 年
竣工时间：2007 年
设计阶段：方案设计、初步设计、施工图设计
建设单位：山东省博兴县人民政府
基地面积：28 000m^2
建筑面积：30 000m^2
主要结构：钢筋混凝土框架结构
主要用途：办公

Project：Administration Building of Boxing County, Shandong
Construction Site：Boxing County, Shandong
Design Period：2005
Completion：2007
Design Phase：Concept Design, Developments Design, Construction Documents Design
Client：Boxing County People's Government, Shandong
Site Area：28,000m^2
Floor Area：30,000m^2
Main Structure：Reinforced Concrete Frame Structure
Main Application：Office

项目名称：嘉定区法院、人民检察院、公安分局业务用房
建设地点：上海市嘉定区
设计时间：2006 年
竣工时间：2011 年
设计阶段：方案设计、初步设计、施工图设计
建设单位：上海市嘉定区机关事务管理局
基地面积：46 182.7m^2
建筑面积：95 736.8m^2
主要结构：框架剪力墙结构
主要用途：办公

Project：The New Office Buildings for the Public Security Bureau, the Court and the Public Prosecutor Mansion of Jiading District
Construction Site：Jiading District, Shanghai
Design Period：2006
Completion：2011
Design Phase：Concept Design, Development Design, Construction Documents Design
Client： Jiading District Bureau of administration, Shanghai
Site Area：46,182.7m^2
Floor Area：95,736.8m^2
Main Structure：Frame shear structure
Main Appliction：Office

嘉定区法院、人民检察院、公安分局业务用房

The New Office Buildings for the Public Security Bureau, the Court and the Public Prosecutor Mansion of Jiading District

武钢技术中心科技大厦

The Science & Technology Building of Wuhan Steel Technical Center

项目名称：武钢技术中心科技大厦	Project：The Science & Technology Building of Wuhan Steel Technical Center
建设地点：湖北省武汉市	Construction Site：Wuhan, Hubei
设计时间：2004 年	Design Period：2004
竣工时间：2006 年	Completion：2006
设计阶段：方案设计、扩初设计、施工图设计	Design Phase：Concept Design, Development Design, Construction Documents Design
建设单位：武汉钢铁（集团）公司	Client: Wuhan Iron and Steel（Gruop）Corp
基地面积：62 047m²	Site Area：62,047m²
建筑面积：41 569m²	Floor Area：41,569m²
主要结构：钢筋混凝土框架结构、钢结构	Main Structure：Reinforced Concrete Frame Structure, Steel Frame Structure
主要用途：办公	Main Application：Office

葛洲坝大厦
GC Tower

项目名称：葛洲坝大厦

建筑地点：上海浦东新区

设计时间：2005—2006 年

竣工时间：2008 年

设计阶段：方案设计、初步设计、施工图设计

建设单位：上海葛洲坝阳明置业有限公司

基地面积：8 174m²

建筑面积：60 706m²

主要结构：钢筋混凝土框架核心筒结构

主要用途：办公

Project：GC Tower

Construction Site：Pudong New District, Shanghai

Design Period：2005-2006

Completion：2008

Design Phase：Concept Design, Development Design, Construction Documents Design

Client：Shanghai Gezhouba Yangming Co.,Ltd.

Site Area：8,174m²

Floor Area：60,706m²

Main Structure：Reinforced Concrete Frame Core Tube Structure

Main Application：Office

上海东方体育中心　新闻服务中心
Press Center, Shanghai Oriental Sports Center

项目名称：上海东方体育中心 新闻服务中心
建筑地点：上海市浦东新区
设计时间：2009—2010 年
竣工时间：2011 年
设计阶段：初步设计、施工图设计
合作单位：德国 gmp 国际建筑设计有限公司
建设单位：上海市体育局
基地面积：259 426m^2
建筑面积：24 230m^2
主要结构：钢筋混凝土框架结构、钢结构
主要用途：体育建筑

Project：Press Center, Shanghai Oriental Sports Center
Construction Site：Pudong New District, Shanghai
Design Period：2009-2010
Completion：2011
Design Phase：Developments Design, Construction Documents Design
Partnership：gmp. Architekten von Gerkan, Marg und Partner
Client：Shanghai Municipal Physcial Culture Bureau
Site Area：259,426m^2
Floor Area：24,230m^2
Main Structure：Reinforced Concrete Frame Structure, Steel Frame Structure
Main Application：Sports Building

波司登国际总部大厦
Bosodeng Headquarters Office Complex, Shanghai

项目名称：波司登国际总部大厦
建筑地点：上海市
设计时间：2008 年
竣工时间：2012 年
设计阶段：方案设计、初步设计、施工图设计
建设单位：上海波司登投资发展有限公司
基地面积：10 311m²
建筑面积：60 880m²
主要结构：钢筋混凝框筒结构
主要用途：办公、商业

Project：Bosideng Headquarters Office Complex, Shanghai
Construction Site：Shanghai
Design Period：2008
Completion：2012
Design Phase：Concept Design, Development Design, Construction Documents Design
Client：Shanghai Bosideng Investment Development Co., Ltd
Site Area：10 ,311m²
Floor Area：60 ,880m²
Main Structure：Reinforced Concrete Frame Main Structure
Main Application：Office Complex

大连检测科技园
Examing Technology Park, Dalian

项目名称：大连检测科技园
建设地点：辽宁省大连市
设计时间：2009 年
竣工时间：2014 年
设计阶段：方案设计、初步设计、施工图设计
建设单位：大连标准检测技术研究中心
基地面积：75 731.2m^2
建筑面积：90 650.3m^2
主要结构：钢筋混凝土框架
主要用途：办公

Project：Examing Technology Park, Dalian
Construction Site：Dalian, Liaoning
Design Period：2009
Completion：2014
Design Phase：Concept Design, Development Design, Construction Documents Design
Client：Dalian Standard Examing Technology Research Center
Site Area：75,731.2m^2
Floor Area：90,650.3m^2
Main Structure：Reinforced Concrete Frame Structure
Main Application：Office

同济规划大厦
Tongji Planning Building

项目名称：同济规划大厦
建设地点：上海市杨浦区
设计时间：2006 年
竣工时间：2009 年
设计阶段：方案设计、初步设计、施工图设计
建设单位：上海市鼎世置业有限公司
基地面积：2 229m^2
建筑面积：24 622m^2
主要结构：框架抗震墙
主要用途：办公

Project：Tongji Planning Building
Construction Site：Yangpu District, Shanghai
Design Period：2006
Completion：2009
Design Phase：Concept Design, Development Design, Construction Documents Design
Client：Shanghai Dingshi Property Co., Ltd.
Site Area：2,229m^2
Floor Area：24,622m^2
Main Structure：Frame Aseismic Wall
Main Appliction：Office

美奂大酒店

Meihuan Grand Hotel

项目名称：美奂大酒店
建筑地点：上海市徐汇区
设计时间：2011—2013 年
竣工时间：2015 年
设计阶段：方案设计、初步设计、施工图设计
建设单位：上海美奂置业有限公司
基地面积：7 004.6m^2
建筑面积：37 926.4m^2
主要结构：钢筋混凝土框架结构
主要用途：酒店

Project：Meihuan Grand Hotel
Construction Site：Xuhui District, Shanghai
Design Period：2011-2013
Completion：2015
Design Phase：Concept Design, Development Design, Construction Documents Design
Partnership：Shanghai Meihuan Real Estate Co. Ltd.
Site Area：7,004.6m^2
Floor Area：37,926.4m^2
Main Structure：Reinforced Concrete Frame Structure
Main Application：Hotel

安徽省稻香楼宾馆徽苑
Huiyuan Guest House in Daoxianglou Hotel, Anhui

项目名称：安徽省稻香楼宾馆徽苑
建筑地点：安徽省合肥
设计时间：2008—2010 年
竣工时间：2010 年
设计阶段：方案设计、景观设计、室内设计
合作单位：安徽省建筑设计研究院
建设单位：安徽省稻香楼宾馆
基地面积：28 000m^2
建筑面积：18 900m^2
主要结构：钢筋混凝土框架结构
主要用途：酒店

Project：Huiyuan Guest House in Daoxianglou Hotel, Anhui
Construction Site：Hefei, Anhui
Design Period：2008-2010
Completion：2010
Design Phase：Scheme Design, Landscape Design, Interior Design
Partnership：Anhui Provincial Architectural Design and Research Institute
Client：Daoxianglou Hotel of Anhui Province
Site Area：28,000m^2
Floor Area：18,900m^2
Main Structure：Reinforced Concrete Frame Structure
Main Application：Hotel

安徽省稻香楼宾馆桂苑
Guiyuan Guest House in Daoxianglou Hotel, Anhui

项目名称：安徽省稻香楼宾馆桂苑
建筑地点：安徽合肥
设计时间：2005—2007 年
竣工时间：2007 年
设计阶段：方案设计、景观设计、室内设计
合作单位：安徽省建筑设计研究院
建设单位：安徽省稻香楼宾馆
基地面积：17 400m^2
建筑面积：14 400m^2
主要结构：混凝土框架结构
主要用途：酒店

Project：Guiyuan Guest House in Daoxianglou Hotel, Anhui
Construction Site：Hefei, Anhui Province
Design Period：2005-2007
Completion：2007
Design Phase：Scheme Design, Landscape Design, Interior Design
Partnership：Anhui Provincial Architectural Design and Research Institute
Client：Daoxianglou Hotel of Anhui Province
Site Area：17,400m^2
Floor Area：14,400m^2
Main Structure：Reinforced Concrete Frame Structure
Main Application：Hotel

项目名称：金豫商厦
建筑地点：上海市黄浦区
设计时间：2004 年
竣工时间：2008 年
设计阶段：方案设计、初步设计、施工图设计
建设单位：上海金豫置业有限公司
基地面积：6 214m²
建筑面积：28 424m²
主要结构：钢筋混凝土框架结构、钢结构
主要用途：商业

Project：Jinyu Commercial Center
Construction Site：Huangpu District, Shanghai
Design Period：2004
Completion：2008
Design Phase：Concept design、Developments design、Construction documents design
Client：Shanghai Jinyu Real Estate Co. Ltd.
Site Area：6,214m²
Floor Area：28,424m²
Main Structure：Reinforced Concrete Frame Structure, Steel Frame Structure
Main Application：Commercial Building

金豫商厦
Jinyu Commercial Center

项目名称：星巴克世博会最佳实践区特别店
建设地点：上海市 2010 世博会城市最佳实践区
设计时间：2012 年
竣工时间：2014 年
设计阶段：方案设计、初步设计、施工图设计
建设单位：上海世博发展集团
基地面积：800m²
建筑面积：5 000m²
主要结构：钢结构
主要用途：商业

Project：Starbucks Coffee at UBPA
Construction Site：Urban Best Practices Area of 2010 EXPO, Shanghai, China
Design Period：2012
Completion：2014
Design Phase：Concept Design, Development Design, Construction Documents Design
Client：EXPO Shanghai Gruop
Site Area：800m²
Floor Area：5,000m²
Main Structure：Steel Structure
Main Appliction：Commercial Building

星巴克世博会最佳实践区特别店
Starbucks Coffee at UBPA

TARBUCKS COFFEE

昆明老街一期商业项目（正义坊）

Kungming Old Town Street First-stage Commercial Project

项目名称：昆明老街一期商业项目（正义坊）
建筑地点：云南省昆明市
设计时间：2005—2006 年
竣工时间：2009 年
设计阶段：方案设计
建设单位：昆明市之江置业有限公司
基地面积：24 899m^2
建筑面积：92 309m^2
主要结构：钢筋混凝土框架结构
主要用途：商业

Project：Kungming Old Town Street First-stage Commercial Project
Construction Site：Kunming,Yunnan
Design Period：2005-2006
Completion：2009
Design Phase：Concept Design
Partnership：Kunming Zhijiang Real Estate Co. Ltd.
Site Area：24,899m^2
Floor Area：92,309m^2
Main Structure：Reinforced Concrete Frame Structure
Main Appliction：Commercial Building

项目名称：淄博市周村古商城汇龙街区
建筑地点：淄博市周村区
设计时间：2005 年
竣工时间：2007 年
设计阶段：修建性详细规划、方案设计
建设单位：淄博市周村区旅游局
基地面积：58 600m^2
建筑面积：35 400m^2
主要结构：钢筋混凝土框架结构
主要用途：商业

Project：Huilong Quarter by Zhoucun Ancient Commercial Town, Zibo
Construction Site：Zhoucun District, Zibo
Design Period：2005
Completion：2007
Design Phase：Detailed Planning, Concept Design
Client：Zhoucun District Bureau of Tourism, Zibo
Site Area：58,600m^2
Floor Area：35,400m^2
Main Structure：Reinforced Concrete Frame Structure
Main Application：Commercial Building

淄博市周村古商城汇龙街区
Huilong Quarter by Zhoucun Ancient Commercial Town, Zibo

汶川映秀镇商业街
Yinxiu Commercial Street, Wen chuan

项目名称：汶川映秀镇商街
建筑地点：汶川县映秀镇
设计时间：2009 年
竣工时间：2010 年
设计阶段：方案设计
建设单位：汶川县映秀镇镇政府
基地面积：0.69hm²
建筑面积：11 293m²
主要结构：钢筋混凝土框架结构
主要用途：商业

Project：Yinxiu Commercial Street, Wenchuan
Construction Site：Yinxiu Town , Wenchuan
Design Period：2009
Completion：2010
Design Phase：Concept Design
Client：Yinxiu Town People's Government, Wenchuan
Site Area：0.69hm²
Floor Area：11 293m²
Main Structure：Reinforced Concrete Frame Structure
Main Application：Commercial Building

项目名称：云南省第二人民医院门急诊综合楼
建筑地点：云南省昆明市
设计时间：2009 年
竣工时间：2011 年
设计阶段：初步设计
合作单位：云南省建筑设计研究院
建设单位：云南省第二人民医院
基地面积：8 800m²
建筑面积：34 100m²
主要结构：钢筋混凝土框架结构
主要用途：医疗

Project：Complex Building for Outpatient and Emergency Services of the Second Hospital, Yunnan
Construction Site：Kunmin, Yunnan
Design Period：2009
Completion：2011
Design Phase：Developments Design
Partnership：Architecture Design & Research Institute Yunnan Province
Client：The Second Hospital of Yunnan
Site Area：8,800m²
Floor Area：34,100m²
Main Structure：Reinforced Concrete Frame Structure
Main Application：Medical Treatment

云南省第二人民医院门急诊综合楼
Complex Building for Outpatient and Emergency Services of the Second Hospital, Yunnan

自行车

项目名称：广珠城际快速轨道交通工程珠海站
建筑地点：广东省珠海市
设计时间：2006 年
竣工时间：2013 年
设计阶段：方案设计、初步设计、施工图设计
建设单位：广铁集团
建筑面积：90 718m^2（含地下室面积）
主要结构：钢筋混凝土框架结构、钢结构
主要用途：交通

Project：Inter-City Train Station between Guangzhou and Zhuhai, Zhuhai
Construction Site：Zhuhai, Guangdong
Design Period：2006
Completion：2013
Design Phase：Concept Design, Development Design, Construction Documents Design
Client：Guangzhou Railway Group
Floor Area：90,718m^2
Main Structure：Reinforced Concrete Frame Structure, Steel Frame Structure
Main Application：Traffic

广珠城际快速轨道交通工程珠海站
Inter-City Train Station between Guangzhou and Zhuhai, Zhuhai

黄河口生态旅游区游船码头

Cruise Ship Wharf of Yellow River Estuary Ecological Tourism Zone

项目名称：黄河口生态旅游区游船码头
建筑地点：山东省东营市
设计时间：2008—2009 年
竣工时间：2011 年
设计阶段：方案设计、初步设计、施工图设计
建设单位：东营市旅游开发有限公司
基地面积：119 165m^2
建筑面积：8 450m^2
主要结构：悬挂结构体系
主要用途：交通

Project：Cruise Ship Wharf of Yellow River Estuary Ecological Tourism Zone
Construction Site：Dongying, Shandong
Design Period：2008-2009
Completion：2011
Design Phase：Concept Design, Development Design, Construction Documents Design
Client：Dongying Tourism Development Co.,Ltd.
Site Area：119,165m^2
Floor Area：84,50m^2
Main Structure：The Pillar Hanging Structure System
Main Application：Traffic

Beijing 2008

Beijing 2008

北京大学体育馆
Peking University Gymnasium

项目名称：北京大学体育馆
建筑地点：北京市海淀区
设计时间：2005—2006 年
竣工时间：2007 年
设计阶段：方案设计、初步设计、施工图设计
建设单位：北京大学基建工程部
基地面积：20 543m^2
建筑面积：26 525m^2
主要结构：钢筋混凝土框架结构、钢结构
主要用途：体育

Project：Peking University Gymnasium
Construction Site：Haidian District, Beijing
Design Period：2005-2006
Completion：2007
Design Phase：Concept Design, Developments Design, Construction Documents Design
Client：Peking University Infrastructure Project Department
Site Area：20,543m^2
Floor Area：26,525m^2
Main Structure：Reinforced Concrete Frame Structure, Steel Frame Structure
Main Application：Sports Building

上海东方体育中心　室外跳水池
Outdoor Diving Pool, Shanghai Oriental Sports Center

项目名称：上海东方体育中心　室外跳水池
建筑地点：上海市浦东新区
设计时间：2009—2010 年
竣工时间：2011 年
设计阶段：初步设计、施工图设计
合作单位：德国 gmp 国际建筑设计有限公司
建设单位：上海市体育局
基地面积：88 053m²
建筑面积：10 515m²
主要结构：钢筋混凝土框架结构、钢结构
主要用途：体育

Project：Outdoor Diving Pool, Shanghai Oriental Sports Center
Construction Site：Pudong New District, Shanghai
Design Period：2009-2010
Completion：2011
Design Phase：Developments Design, Construction Documents Design
Partnership：gmp.Architekten von Gerkan, Marg und Partner
Client：Shanghai Municipal Physcial Culture Bureau
Site Area：88,053m²
Floor Area：10,515m²
Main Structure：Reinforced Concrete Frame Structure, Steel Frame Structure
Main Application：Sports Building

遂宁体育中心
Suining Sports Center

项目名称：遂宁体育中心
建筑地点：四川省遂宁市
设计时间：2010—2011 年
竣工时间：2014 年
设计阶段：方案设计、初步设计、施工图设计
建设单位：遂宁市河东开发建设投资有限公司
基地面积：127 525m^2
建筑面积：79 190m^2
主要结构：钢筋混凝土框架结构、钢结构
主要用途：体育

Project：Suining Sports Center
Construction Site：Suining, Sichuan
Design Period：2010-2011
Completion：2014
Design Phase：Concept Design, Developments Design, Construction Documents Design
Client：Suining Hedong Development and Construction Investment Co., Ltd.
Site Area：127,525m^2
Floor Area：79,190m^2
Main Structure：Reinforced Concrete Frame Structure, Steel Frame Structure
Main Application：Sports Building

连云港体育中心
Lianyugang Sports Center

项目名称：连云港体育中心
建筑地点：江苏省连云港市
设计时间：2006—2007 年
竣工时间：2008 年
设计阶段：方案设计、初步设计、施工图设计
建设单位：连云港市体育局
基地面积：317 055m^2
建筑面积：74 227m^2
主要结构：钢筋混凝土框架结构、钢结构
主要用途：体育

Project：Lianyungang Sports Center
Construction Site：Lianyungang, Jiangsu
Design Period：2006-2007
Completion：2008
Design Phase：Concept Design, Developments Design, Construction Documents Design
Client：Lianyungang Municipal Physcial Culture Bureau
Site Area：317,055m^2
Floor Area：74,227m^2
Main Structure：Reinforced Concrete Frame Structure, Steel Frame Structure
Main Application：Sports Building

莆田体育中心
Putian Sports Center

项目名称：莆田体育中心
建筑地点：福建省莆田市
设计时间：2006—2007 年
竣工时间：2009 年
设计阶段：方案设计、初步设计、施工图设计
建设单位：莆田市体育局
基地面积：49 500m²
建筑面积：33 016m²
主要结构：钢筋混凝土框架结构、钢结构
主要用途：体育

Project：Putian Sports Center
Construction Site：Putian, Fujian
Design Period：2006-2007
Completion：2009
Design Phase：Concept Design, Developments Design, Construction Documents Design
Client：Putian Municipal Physcial Culture Bureau
Site Area：49,500m²
Floor Area：33,016m²
Main Structure：Reinforced Concrete Frame Structure, Steel Frame Structure
Main Application：Sports Building

南通体育会展中心体育场
Nantong Stadium of Sports Convention and Exhibition Center

项目名称：南通体育会展中心体育场
建筑地点：江苏省南通市
设计时间：2003—2004 年
竣工时间：2005 年
设计阶段：方案设计、初步设计、施工图设计
建设单位：南通市规划管理局
基地面积：400 000m^2
建筑面积：48 600m^2
主要结构：钢筋混凝土框架结构、钢结构
主要用途：体育

Project：Nantong Stadium of Sports Convention and Exhibition Center
Construction Site：Nantong, Jiangsu
Design Period：2003-2004
Completion：2005
Design Phase：Concept Design , Developments Design , Construction Documents Design
Client：Nantong Municipal Planning Bureau
Site Area：400,000m^2
Floor Area：48,600m^2
Main Structure：Reinforced Concrete Frame Structure, Steel Frame Structure
Main Application：Sports Building

盱眙体育中心体育场
Stadium, Xuyi Sports Center

项目名称：盱眙体育中心体育场
建筑地点：江苏省盱眙市
设计时间：2007—2008 年
竣工时间：2009 年
设计阶段：方案设计、初步设计、施工图设计
建设单位：盱眙县人民政府
基地面积：153 341m²
建筑面积：42 234m²
主要结构：钢筋混凝土框架结构、钢结构
主要用途：体育

Project：Stadium, Xuyi Sports Center
Construction Site：Xuyi, Jiangsu
Design Period：2007-2008
Completion：2009
Design Phase：Concept Design, Developments Design, Construction Documents Design
Client：Municipal Government of Xuyi County
Site Area：153,341m²
Floor Area：42,234m²
Main Structure：Reinforced Concrete Frame Structure, Steel Frame Structure
Main Application：Sports Building

西门

济宁奥体中心　体育场
The Stadium, Jining Olympic Sports Center

项目名称：济宁奥体中心　体育场
建筑地点：山东省济宁市
设计时间：2011—2012 年
竣工时间：2014 年
设计阶段：初步设计、施工图设计
合作单位：济宁市文本中心筹建处
建设单位：德国罗昂建筑设计咨询有限公司
基地面积：499 086m^2
建筑面积：33 181m^2
主要结构：钢筋混凝土框架结构、钢结构
主要用途：体育

Project：The Stadium, Jining Olympic Sports Center
Construction Site：Jining, Shandong
Design Period：2011-2012
Completion：2014
Design Phase：Developments Design, Construction Documents Design
Partnership：Jining Preparation Department of Recreation and Sports Center
Client：Logon Urban Architecture Design
Site Area：499,086m^2
Floor Area：33,181m^2
Main Structure：Reinforced Concrete Frame Structure, Steel Frame Structure
Main Application：Sports Building

寿光体育场
Shouguang Stadium

项目名称：寿光体育场
建筑地点：山东省寿光市
设计时间：2007—2008 年
竣工时间：2009 年
设计阶段：方案设计、初步设计、施工图设计
建设单位：寿光市城市基础设施建设投资管理中心
基地面积：491 527m^2
建筑面积：46 972m^2
主要结构：钢筋混凝土框架结构、钢结构
主要用途：体育

Project：Shouguang Stadium
Construction Site：Shouguang, Shandong Provience
Design Period：2007-2008
Completion：2009
Design Phase：Concept Design, Developments Design, Construction Documents Design
Client：Shouguang Urban Infrastrature Investment and Construction Administration Center
Site Area：491,527m^2
Floor Area：46,972m^2
Main Structure：Reinforced Concrete Frame Structure, Steel Frame Structure
Main Application：Sports Building

泉州海峡体育中心体育馆
Gymnasium, Quanzhou Strait Sports Center

项目名称：泉州海峡体育中心体育馆
建筑地点：福建省泉州市
设计时间：2005—2006 年
竣工时间：2008 年
设计阶段：方案设计、初步设计、施工图设计
建设单位：泉州市体育局
基地面积：235 012m^2
建筑面积：29 200m^2
主要结构：钢筋混凝土框架结构、钢结构
主要用途：体育

Project：Gymnasium, Quanzhou Strait Sports Center
Construction Site：Quanzhou, Fujian
Design Period：2005-2006
Completion：2008
Design Phase：Concept Design, Developments Design, Construction Documents Design
Client：Quanzhou Municipal Physcial Culture Bureau
Site Area：235,012m^2
Floor Area：29,200m^2
Main Structure：Reinforced Concrete Frame Structure, Steel Frame Structure
Main Application：Sports Building

常熟市体育中心体育馆
Stadium, Changshu Sports Center

项目名称：常熟市体育中心体育馆
建筑地点：江苏省常熟市
设计时间：2009—2010 年
竣工时间：2012 年
设计阶段：方案设计、初步设计、施工图设计
建设单位：常熟市城市经营投资有限公司
基地面积：110 467m^2
建筑面积：32 249m^2
主要结构：钢筋混凝土框架结构、钢结构
主要用途：体育

Project：Stadium, Changshu Sports Center
Construction Site：Changshu, Jiangsu
Design Period：2009-2010
Completion：2012
Design Phase：Concept Design, Developments Design, Construction Documents Design
Client：Changshu Urban Investment Co., Ltd.
Site Area：110,467m^2
Floor Area：32,249m^2
Main Structure：Reinforced Concrete Frame Structure, Steel Frame Structure
Main Application：Sports Building

济宁奥体中心体育馆
The Gymnasium Jining Olympic Sports Center

项目名称：济宁奥体中心 体育馆
建筑地点：山东省济宁市
设计时间：2011—2012 年
竣工时间：2014 年
设计阶段：初步设计、施工图设计
合作单位：斯构莫尼建筑设计咨询（上海）有限公司
建设单位：济宁市文体中心筹建处
基地面积：499 086m^2
建筑面积：40 993m^2
主要结构：钢筋混凝土框架结构、钢结构
主要用途：体育

Project：The Gymnasium, Jining Olympic Sports Center
Construction Site：Jining, Shandong
Design Period：2011-2012
Completion：2014
Design Phase：Developments Design, Construction Documents Design
Partnership：Studio Consociato Architettura Urbanistica
Client：Jining Preparation Department of Recreation and Sports Center
Site Area：499,086m^2
Floor Area：40,993m^2
Main Structure：Reinforced Concrete Frame Structure, Steel Frame Structure
Main Application：Sports Building

东风日产

济宁奥体中心游泳馆

The Natatorium Jining Olympic Sports Center

项目名称：济宁奥体中心游泳馆

建筑地点：山东省济宁市

设计时间：2011—2012 年

竣工时间：2014 年

设计阶段：初步设计、施工图设计

合作单位：斯构莫尼建筑设计咨询（上海）有限公司

建设单位：济宁市文体中心筹建处

基地面积：499 086m^2

建筑面积：32 167m^2

主要结构：钢筋混凝土框架结构、钢结构

主要用途：体育

Project：The Natatorium, Jining Olympic Sports Center

Construction Site：Jining, Shandong

Design Period：2011-2012

Completion：2014

Design Phase：Developments Design, Construction Documents Design

Partnership：Studio Consociato Architettura Urbanistica

Client：Jining Preparation Department of Recreation and Sports Center

Site Area：499,086m^2

Floor Area：32,167m^2

Main Structure：Reinforced Concrete Frame Structure, Steel Frame Structure

Main Application：Sports Building

日照游泳馆
Rizhao Natatorium

项目名称：日照游泳馆
建筑地点：山东省日照市
设计时间：2008—2009 年
竣工时间：2010 年
设计阶段：方案设计、初步设计、施工图设计
建设单位：日照市规划建设委员会
基地面积：412 410m²
建筑面积：29 220m²
主要结构：钢筋混凝土框架结构、钢结构
主要用途：体育

Project：Rizhao Natatorium
Construction Site：Rizhao, Shandong
Design Period：2008-2009
Completion：2010
Design Phase：Concept Design, Developments Design, Construction Documents Design
Client：Rizhao Planning and Construction Bureau
Site Area：412,410m²
Floor Area：29,220m²
Main Structure：Reinforced Concrete Frame Structure, Steel Frame Structure
Main Application：Sports Building

东4
东3

济宁奥体中心射击馆
The Shooting Hall, Jining Olympic Sports Center

项目名称：济宁奥体中心射击馆
建筑地点：山东省济宁市
设计时间：2011—2012 年
竣工时间：2014 年
设计阶段：初步设计、施工图设计
合作单位：斯构莫尼建筑设计咨询（上海）有限公司
建设单位：济宁市文体中心筹建处
基地面积：499 086m^2
建筑面积：27 916m^2
主要结构：钢筋混凝土框架结构、钢结构
主要用途：体育

Project：The Shooting Hall, Jining Olympic Sports Center
Construction Site：Jining, Shandong Provience
Design Period：2011-2012
Completion：2014
Design Phase：Developments Design, Construction Documents Design
Partnership：Studio Consociato Architettura Urbanistica
Client：Jining Preparation Department of Recreation and Sports Center
Site Area：499,086m^2
Floor Area：27,916m^2
Main Structure：Reinforced Concrete Frame Structure, Steel Frame Structure
Main Application：Sports Building

四川川投国际网球中心
Sichuan International Tennis Center

项目名称：四川川投国际网球中心
建筑地点：四川省成都市
设计时间：2006—2007 年
竣工时间：2008 年
设计阶段：方案设计、初步设计、施工图设计
建设单位：四川川投国际网球中心开发有限责任公司
基地面积：259 530m²
建筑面积：83 185m²
主要结构：钢结构
主要用途：体育

Project：Sichuan International Tennis Center
Construction Site：Chengdu, Sichuan
Design Period：2006-2007
Completion：2008
Design Phase：Concept Design, Development Design, Construction Documents Design
Client：Sichuan International Tennis Center Development Co., Ltd.
Site Area：259,530m²
Floor Area：83,185m²
Main Structure：Steel Structures
Main Application：Sports Building

中国残疾人体育艺术培训基地（诺宝中心）
Sports and Arts Training Base of China Disabled Persons （Nobel Center）

项目名称：中国残疾人体育艺术培训基地（诺宝中心）
建设地点：上海市闵行区
设计时间：2002 年
竣工时间：2005 年
设计阶段：方案设计、初步设计、施工图设计
建设单位：中国残疾人联合会
基地面积：39 906m^2
建筑面积：23 378m^2
主要结构：钢筋混凝土框架结构、钢结构
主要用途：训练、培训

Project：Sports and Arts Training Base of China Disabled Persons（Nobel Center）
Construction Site：Minhang District, Shanghai
Design Period：2002
Completion：2005
Design Phase：Concept Design, Development Design, Construction Documents Design
Client：China Disabled Persons'Federation
Site Area：39,906m^2
Floor Area：23,378m^2
Main Structure：Reinforced Concrete Frame Structure, Steel Frame Structure
Main Application：Training

北川国家地震遗址博物馆

Beichuan National Earthquake Ruins Museum

项目名称：北川国家地震遗址博物馆
建筑地点：绵阳市北川县
设计时间：2008 年
竣工时间：2010 年
设计阶段：整体方案策划、保护区规划与设计
合作单位：上海同济城市规划设计研究院
建设单位：绵阳市人民政府
控制区面积：657.38hm^2
保护区面积：74.27hm^2
主要用途：纪念性建筑

Project：Beichuan National Earthquake Ruins Museum
Construction Site：Beichuan County, Mianyang
Design Period：2008
Completion：2010
Design Phase：Comprehensive Programme planning, Planning and Design for Preservation Zone
Partnership：Shanghai Tongji Urban Planning and Research Institute
Client：Mianyang Municipal People's Government
Area of Control Zone：657.38hm^2
Area of Preservation Zone：74.27hm^2
Main Application：Memorial Building

都江堰市壹街区图书馆
Public Library, Dujiangyan

项目名称：都江堰市图书馆
建筑地点：四川省都江堰市
设计时间：2008 年
竣工时间：2010 年
设计阶段：方案设计、初步设计、施工图设计
建设单位：都江堰新城建设投资有限公司
基地面积：3 800 m²
建筑面积：5 500 m²
主要结构：钢筋混凝土框架结构
主要用途：文化

Project：Public Library, Dujiangyan
Construction Site：Dujiangyan, Sichuan
Design Period：2008
Completion：2010
Design Phase：Concept Design, Development Design, Construction Documents Design
Client：Dujiangyan New City Construction Investment Co., Ltd.
Site Area：3,800m²
Floor Area：5,500m²
Main Structure：Reinforced Concrete Frame Structure
Main Application：Culture

西藏日喀则桑珠孜宗堡（宗山城堡）保存与再生工程
Restoration and Regeneration of Sangzhutse Fortress in Shigatse, Tibet

项目名称：西藏日喀则桑珠孜宗堡（宗山城堡）保存与再生工程
建筑地点：西藏日喀则市
设计时间：2004 年
竣工时间：2009 年
设计阶段：方案设计、初步设计、施工图设计
建设单位：上海市人民政府援藏援疆领导小组办公室
基地面积：44 000m^2
建筑面积：8 326m^2
主要结构：钢筋混凝土框架结构
主要用途：博物馆

Project：Restoration and Regeneration of Sangzhutse Fortress in Shigatse, Tibet
Construction Site：Shigatse, Tibet
Design Period：2004
Completion：2009
Design Phase：Concept Design, Developments Design, Construction Documents Design
Client：Shanghai's Financial Aided Program Office for Tibet and Xinjiang
Site Area：44,000m^2
Floor Area：8,326m^2
Main Structure：Reinforced Concrete Frame Structure
Main Application：Museum

上海徐汇风貌保护道路规划

Detailed Planning Guidelines for Historical Streets in Xuhui District, Shanghai

项目名称：上海徐汇风貌保护道路规划
项目地点：上海市徐汇区
编制时间：2011—2013 年
规划阶段：修建性详细规划
组织编制单位：徐汇区规划和土地管理局
规划控制面积：420hm^2
保护区范围：40 条街道

Project：Detailed Planning Guidelines for Historical Streets in Xuhui District, Shanghai
Project Site：Xuhui District, Shanghai
Planning Period：2011-2013
Design Phase：Detailed Planning
Client：Shanghai Xuhui District Planning and Land Administration Bureau
Area of Control Zone：420hm^2
Area of Preservation Zone：40 Streets

项目名称：南市发电厂主厂房和烟囱改造工程——城市未来馆
建设地点：上海市 2010 世博会城市最佳实践区
设计时间：2009 年
竣工时间：2010 年
设计阶段：方案设计、初步设计、施工图设计
建设单位：上海世博土地控股有限公司
基地面积：19 103m^2
建筑面积：31 088m^2
主要结构：钢筋混凝土框架结构、钢结构
主要用途：展览馆

Project：Renovation project of Nanshi Power Plant and Chimney (City Future Pavilion)
Construction Site：Urban Best Practices Area of 2010 EXPO, Shanghai, China
Design Period：2009
Completion：2010
Design Phase：Concept Design, Development Design, Construction Documents Design
Client：Shanghai World Expo Land Holdings Co., Ltd.
Site Area：19,103m^2
Floor Area：31,088m^2
Main Structure：Reinforced Concrete Frame Structure, Steel Frame Structure
Main Appliction：Exhibition Hall

南市发电厂主厂房和烟囱改造工程——城市未来馆
Renovation Project of Nanshi Power Plant and Chimney（City Future Pavilion）

轮船招商总局大楼修缮工程
Regeneration Project of Merchant Steamship Bureau Building, Shanghai

项目名称：轮船招商总局大楼修缮工程
建筑地点：上海市
设计时间：2002 年
竣工时间：2004 年
设计阶段：方案设计、初步设计、施工图设计
建设单位：招商局集团（上海）有限公司
基地面积：654m^2
建筑面积：1 672.5m^2
主要结构：钢筋混凝土框架结构
主要用途：办公

Project：Regeneration Project of Merchant Steamship Bureau Building, Shanghai
Construction Site：Shanghai
Design Period：2002
Completion：2004
Design Phase：Concept Design, Developments Design, Construction Documents Design
Client：China Merchants Group (Shanghai) Co., Ltd.
Site Area：654m^2
Floor Area：1,672.5m^2
Main Structure：Reinforced Concrete Frame Structure
Main Application：Office

大华清水湾三期老建筑保护与再生工程

Conservation and Renouation Project of the Historic Building in Qingshuiwan Residential Area

项目名称：大华清水湾三期老建筑保护与再生工程
建筑地点：上海市
设计时间：2008 年
竣工时间：2011 年
设计阶段：方案设计、初步设计、施工图设计
建设单位：上海华运房地产开发有限公司
基地面积：618m^2
建筑面积：2 424.1m^2
主要结构：钢筋混凝土框架结构和钢结构
主要用途：商业

Project：Conservation and Renouation Project of the Historic Building in Qingshuiwan Residential Area
Construction Site：Shanghai
Design Period：2008
Completion：2011
Design Phase：Concept Design, Developments Design, Construction Documents Design
Client：Shanghai Huayun Real Estate Development Co. Ltd.
Site Area：618m^2
Floor Area：2,424.1m^2
Main Structure：Reinforced Concrete Frame Structure and Steel Structure
Main Application：Commercial Building

上海啤酒公司建筑修缮工程
Renovation Project of the Workshops in Union Brewery Ltd., Shanghai

项目名称：上海啤酒公司建筑修缮工程
建筑地点：上海市普陀区
设计时间：2003—2004 年
竣工时间：2005 年
设计阶段：方案设计、初步设计
合作单位：上海房屋建筑设计研究院
建设单位：上海市苏州河综合整治建设有限公司
基地面积：35 500m^2
建筑面积：改造前 18 300m^2；改造后 9 228m^2
主要结构：钢筋混凝土框架结构
主要用途：展示、商业

Project：Renovation Project of the Workshops in Union Brewery Ltd., Shanghai
Construction site：Putuo District, Shanghai
Design period：2003-2004
Completion：2005
Design Phase：Scheme Design, Design Development
Partnership：Shanghai Municipal Housing Design Institute
Client：Shanghai Suzhou Creek Comprehensive Improvement Construction Co. Ltd.
Site area：35,500m^2
Floor area：Before Renovation：18,300m^2; After Renovation：9,228m^2
Main Structure：Reinforced Concrete Frame Structure
Main Application：Exhibition & Commercial

项目名称：世博会城市最佳实践区中部展馆 B–3 馆
建筑地点：上海市 2010 世博会城市最佳实践区
设计时间：2008 年
竣工时间：2010 年
设计阶段：方案设计、初步设计、施工图设计
建设单位：上海世博会事务协调局
基地面积：14 390m²
建筑面积：8 500m²
主要结构：钢结构
主要用途：展览

Project：Expo UBPA Central Pavilion B-3
Construction Site：Urban Best Practices Area of 2010 EXPO, Shanghai, China
Design Period：2008
Completion：2010
Design Phase：Concept Design, Development Design, Construction Documents Design
Client：Coordination Bureau of Shanghai World Expo
Site Area：14,390m²
Floor Area：8,500m²
Main Structure：Steel Structures
Main Application：Exhibition

世博会城市最佳实践区中部展馆 B–3 馆
Expo UBPA Central Pavilion B-3

同济大学一・二九大楼改建工程（同济博物馆）
Renovation Project of Building "1·29", Tongji University

项目名称：同济大学一・二九大楼改建工程（同济博物馆）
建筑地点：同济大学
设计时间：2010—2012 年
竣工时间：2013 年
设计阶段：方案设计、初步设计、施工图设计
建设单位：同济大学
基地面积：3 560m^2
建筑面积：4 469m^2
主要结构：砖木混合结构
主要用途：博物馆

Project：Renovation Project of Building "1·29",
Tongji University, Shanghai
Construction Site：Campus of Tongji University, Shanghai
Design Period：2010-2012
Completion：2013
Design Phase：Concept Design, Development Design,
Construction Documents Design
Client：Tongji University
Site Area：3,560m^2
Floor Area：4,469m^2
Main Structure：Masonry and Wood Structure
Main Application：Exhibition

同济大学建筑城规学院 D 楼改进工程
Renouation Project of Building D for CAUP Tongji University

项目名称：同济大学建筑城规学院 D 楼改进工程
建筑地点：同济大学
设计时间：2008—2009 年
竣工时间：2010 年
设计阶段：方案设计、初步设计、施工图设计
建设单位：同济大学
基地面积：3 201m²
建筑面积：6 440m²
主要结构：钢筋混凝土框架结构
主要用途：教育

Project：Renouation Project of Building D for CAUP Tongji University, Shanghai
Construction Site：Campus of Tongji University, Shanghai
Design Period：2008-2009
Completion：2010
Design Phase：Concept Design, Development Design, Construction Documents Design
Client：Tongji University
Site Area：3,201m²
Floor Area：6,440m²
Main Structure：Reinforced Concrete Frame Structure
Main Application：Education

同济大礼堂保护性改建
Renovation Project of Tongji University Auditorium

项目名称：同济大礼堂保护性改建
建筑地点：同济大学
设计时间：2005—2006 年
竣工时间：2007 年
设计阶段：方案设计、初步设计、施工图设计
建设单位：同济大学
基地面积：8 500m^2
建筑面积：7 203m^2
主要结构：装配整体式钢筋混凝土联方网架结构
主要用途：会议

Project：Renovation Project of Tongji University Auditorium
Construction Site：Campus of Tongji University, Shanghai
Design Period：2005-2006
Completion：2007
Design Phase：Concept Design, Development Design, Construction Documents Design
Client：Tongji University
Site Area：8,500m^2
Floor Area：7,203m^2
Main Structure：Assembled Monolithic Reinforced Concrete Union Square Space Truss Structure
Main Application：Conference

北站社区文化活动中心
Cultural Center of North Station Community

项目名称：北站社区文化活动中心	Project：Cultural Center of North Station Community
建设地点：上海市闸北区	Construction Site：Zhabei District, Shanghai
设计时间：2006 年	Design Period：2006
竣工时间：2007 年	Completion：2007
设计阶段：方案设计、初步设计、施工图设计	Design Phase：Concept Design, Development Design, Construction Documents Design
建设单位：上海市闸北区北站街道办事处	Client：Zhabei District North Street Office, Shanghai
基地面积：870m^2	Site Area：870m^2
建筑面积：2 500m^2	Floor Area：2,500m^2
主要结构：砖混结构、钢结构	Main Structure：Brick and Concrete Structure, Steel Structure
主要用途：文化	Main Appliction：Culture

项目名称：原作设计工作室改造	Project：Renovation Project of Original Design Studio
建设地点：上海市杨浦区	Construction Site：Yangpu District, Shanghai
设计时间：2013 年	Design Period：2013
竣工时间：2013 年	Completion：2013
设计阶段：方案设计、初步设计、施工图设计	Design Phase：Concept Design, Development Design, Construction Documents Design
建设单位：原作设计工作室	Client：Original Design Studio
基地面积：800m^2	Site Area：800m^2
建筑面积：1 000m^2	Floor Area：1,000m^2
主要结构：砖木结构	Main Structure：Brick-Wood Structure
主要用途：办公	Main Appliction：Office

原作设计工作室改造
Renovation Project of Original Design Studio

江苏省泰州中学老校区保护性改造工程

Renovation Project of Old Campus of Taizhou High School, Jiangsu

项目名称：江苏省泰州中学老校区保护性改造工程
建筑地点：江苏省泰州市
设计时间：2007 年
竣工时间：2010 年
设计阶段：方案设计、初步设计、施工图设计
建设单位：泰州市住房和城乡建设局
基地面积：42 246m²
建筑面积：6 444m²（新建教学楼）
主要结构：钢筋混凝土框架结构
主要用途：教育

Project：Renovation Project of Old Campus of Taizhou High School, Jiangsu Province
Construction Site：Taizhou, Jiangsu
Design Period：2007
Completion：2010
Design Phase：Concept Design, Development Design, Construction Documents Design
Client：Bureau of Housing and Urban-Rural Development of Taizhou
Site Area：42,246m²
Floor Area：6,444m²（New Teaching Building）
Main Structure：Reinforced Concrete Frame Structure
Main Application：Eduction

上实东滩低碳农业园小粮仓室内外环境设计
Small Barn's Interior & Exterior Environmental Design in SIIC Dongtan Low Carbon Agriculture Park

项目名称：上实东滩低碳农业园小粮仓室内外环境设计
建筑地点：上海市崇明县
设计时间：2011 年
竣工时间：2013 年
设计阶段：方案设计、初步设计、施工图设计
建设单位：上海实业东滩投资开发（集团）有限公司
建筑面积：1 300m^2（三幢）
主要结构：钢筋混凝土框架结构
主要用途：办公、展示

Project：Small Barn's Interior & Exterior Environmental Design in SIIC Dongtan Low Carbon Agriculture Park
Construction Site：Chongming County, Shanghai
Design Period：2011
Completion：2013
Design Phase：Concept Design, Development Design, Construction Documents Design
Client：SIIC Dongtan Investment & Development（Holdings）Co., Ltd.
Floor Area：1,300m^2（3 Buildings）
Main Structure：Reinforced Concrete Frame Structure
Main Application：Office, Exhibition

都江堰“壹街区”安居房灾后重建项目
The Project of “No.1 District” Relief Residence, Dujiangyan

项目名称：都江堰“壹街区”安居房灾后重建项目
建筑地点：四川省都江堰市
设计时间：2008 年
竣工时间：2010 年
设计阶段：方案设计、扩初设计、施工图设计
建设单位：都江堰新城建设投资有限公司
基地面积：30hm²
建筑面积：280 000m²
主要结构：钢筋混凝土框架结构
主要用途：住宅

Project：The Project of “No.1 District ” Relief Residence, Dujiangyan
Construction Site：Dujiangyan, Sichuan
Design Period：2008
Completion：2010
Design Phase：Concept Design, Development Design, Construction Documents Design
Client：Dujiangyan New City Construction Investment Co., Ltd.
Site Area：30hm²
Floor Area：280,000m²
Main Structure：Reinforced Concrete Frame Structure
Main Application：Residence

都江堰市“壹街区”安居房灾后重建项目（F10/K07 地块）
“No.1 District” Relief Residence Block F10/K07, Dujiangyan

项目名称：都江堰市“壹街区”安居房灾后重建项目（F10/K07 地块）
建筑地点：四川省都江堰市
设计时间：2008 年
竣工时间：2010 年
设计阶段：方案设计、初步设计、施工图设计
建设单位：都江堰新城建设投资有限公司
基地面积：18 260m^2
建筑面积：41 820m^2
主要结构：钢筋混凝土框架结构
主要用途：住宅

Project：“No.1 District” Relief Residence Block F10/K07, Dujiangyan
Construction Site：Dujiangyan, Sichuan
Design Period：2008
Completion：2010
Design Phase：Concept Design, Development Design, Construction Documents Design
Client：Dujiangyan New City Construction Investment Co., Ltd.
Site Area：18,260m^2
Floor Area：41,820m^2
Main Structure：Reinforced Concrete Frame Structure
Main Application：Residence

交通银行

lenovo
交通银行

都江堰市“壹街区”安居房灾后重建项目（K10/K11 地块）
“No.1 District” Relief Residence Block K10/K11, Dujiangyan

项目名称：都江堰市“壹街区”安居房灾后重建项目（K10/K11 地块）
建筑地点：四川省都江堰市
设计时间：2008 年
竣工时间：2010 年
设计阶段：方案设计、初步设计、施工图设计
建设单位：都江堰新城建设投资有限公司
基地面积：16 240m^2
建筑面积：32 918m^2
主要结构：钢筋混凝土框架结构
主要用途：住宅

Project：“No.1 District” Relief residence Block K10/K11, Dujiangyan
Construction Site：Dujiangyan, Sichuan
Design Period：2008
Completion：2010
Design Phase：Concept Design, Development Design, Construction Documents Design
Client：Dujiangyan New City Construction Investment Co., Ltd.
Site Area：16,240m^2
Floor Area：32,918m^2
Main Structure：Reinforced Concrete Frame Structure
Main Application：Residence

前方学校

预祝海棠园业主委员会成立

都江堰市“壹街区”安居房灾后重建项目（F01/04/07 地块）
“No.1 District” Relief Residence Block F01/04/07, Dujiangyan

项目名称：都江堰市“壹街区”安居房灾后重建项目（F01/04/07 地块）
建筑地点：四川省都江堰市
设计时间：2008 年
竣工时间：2010 年
设计阶段：方案设计、初步设计、施工图设计
建设单位：都江堰新城建设投资有限公司
基地面积：31 350m²
建筑面积：61 210 m²
主要结构：钢筋混凝土框架结构
主要用途：住宅

Project："No.1 District" Relief Residence Block F01/04/07, Dujiangyan
Construction Site：Dujiangyan, Sichuan
Design Period：2008
Completion：2010
Design Phase：Concept Design, Development Design, Construction Documents Design
Client：Dujiangyan New City Construction Investment Co., Ltd.
Site Area：31,350m²
Floor Area：61,210m²
Main Structure：Reinforced Concrete Frame Structure
Main Application：Residence

都江堰市“壹街区”安居房灾后重建项目（C01 地块）
“No.1 District” Relief Residence Block C01, Dujiangyan

项目名称：都江堰市“壹街区”安居房灾后重建项目（C01 地块）
建筑地点：四川省都江堰市
设计时间：2008 年
竣工时间：2010 年
设计阶段：方案设计、初步设计、施工图设计
建设单位：都江堰新城建设投资有限公司
基地面积：14 480m²
建筑面积：24 660 ㎡
主要结构：钢筋混凝土框架结构
主要用途：住宅

Project：“No.1 District” Relief residence Block C01, Dujiangyan
Construction Site：Dujiangyan, Sichuan
Design Period：2008
Completion：2010
Design Phase：Concept Design, Development Design, Construction Documents Design
Client：Dujiangyan New City Construction Investment Co., Ltd.
Site Area：14,480m²
Floor Area：43,654m²
Main Structure：Reinforced Concrete Frame Structure
Main Application：Residence

岷江攝影藝術
MINJIANG PHOTOGRAPHY CENTER

都江堰市“壹街区”安居房灾后重建项目（F02/F05 地块）
“No.1 District” Relief Residence Block F02/F05, Dujiangyan

项目名称：都江堰市“壹街区”安居房灾后重建项目（F02/F05 地块）
建筑地点：四川省都江堰市
设计时间：2008 年
竣工时间：2010 年
设计阶段：方案设计、初步设计、施工图设计
建设单位：都江堰新城建设投资有限公司
基地面积：14 270m²
建筑面积：23 620m²
主要结构：钢筋混凝土框架结构
主要用途：住宅

Project: “No.1 District” Relief Residence Block F02/F05, Dujiangyan
Construction Site: Dujiangyan, Sichuan
Design Period: 2008
Completion: 2010
Design Phase: Concept Design, Development Design, Construction Documents Design
Client: Dujiangyan New City Construction Investment Co., Ltd.
Site Area: 14,270m²
Floor Area: 23,620m²
Main Structure: Reinforced Concrete Frame Structure
Main Application: Residence

都江堰市“壹街区”安居房灾后重建项目（K01/F06 地块）
“No.1 District” Relief Residence Block K01/F06, Dujiangyan

项目名称：都江堰市“壹街区”安居房灾后重建项目（K01/F06 地块）
建筑地点：四川省都江堰市
设计时间：2008 年
竣工时间：2010 年
设计阶段：方案设计、初步设计、施工图设计
建设单位：都江堰新城建设投资有限公司
基地面积：15 830m^2
建筑面积：33 684m^2
主要结构：钢筋混凝土框架结构
主要用途：住宅

Project：“No.1 District” Relief Residence Block K01/F06 , Dujiangyan
Construction Site：Dujiangyan, Sichuan
Design Period：2008
Completion：2010
Design Phase：Concept Design, Development Design, Construction Documents Design
Client：Dujiangyan New City Construction Investment Co., Ltd.
Site Area：15,830m^2
Floor Area：33,684m^2
Main Structure：Reinforced Concrete Frame Structure
Main Application：Residence

KUTOO
酷途户外

酷途户外

都江堰市“壹街区”安居房灾后重建项目（K02/03 地块）
“No.1 District” Relief Residence Block K02/03, Dujiangyan

项目名称：都江堰市“壹街区”安居房灾后重建项目（K02/03 地块）
建筑地点：四川省都江堰市
设计时间：2008 年
竣工时间：2010 年
设计阶段：方案设计、初步设计、施工图设计
建设单位：都江堰新城建设投资有限公司
基地面积：23 780m²
建筑面积：42 000m²
主要结构：钢筋混凝土框架结构
主要用途：住宅

Project：“No.1 District” Relief Residence Block K02/03, Dujiangyan
Construction Site：Dujiangyan, Sichuan
Design Period：2008
Completion：2010
Design Phase：Concept Design, Development Design, Construction Documents Design
Client：Dujiangyan New City Construction Investment Co., Ltd.
Site Area：23,780m²
Floor Area：42,000m²
Main Structure：Reinforced Concrete Frame Structure
Main Application：Residence

项目名称：汶川映秀镇中心 13 号地块民房重建
建筑地点：汶川县映秀镇
设计时间：2008 年
竣工时间：2010 年
设计阶段：方案设计、初步设计、施工图设计
建设单位：映秀镇人民政府
基地面积：8 900m^2
建筑面积：12 000m^2
主要结构：钢筋混凝土框架结构
主要用途：住宅

Project：Yinxiu Relief Residence Block13, Wenchuan
Construction Site：Yinxiu Town , Wenchuan
Design Period：2008
Completion：2010
Design Phase：Concept Design, Development Design, Construction Documents Design
Client：Yinxiu Town People's Government, Wenchuan
Site Area：8,900m^2
Floor Area：12,000m^2
Main Structure：Reinforced Concrete Frame Structure
Main Application：Residence

汶川映秀镇中心 13 号地块民房重建
Yinxiu Relief Residence Block13, Wenchuan

项目名称：汶川映秀镇中心 15 号地块民房重建	Project：Yinxiu Relief Residence Block15, Wenchuan
建筑地点：汶川县映秀镇	Construction Site：Yinxiu Town , Wenchuan
设计时间：2008 年	Design Period：2008
竣工时间：2010 年	Completion：2010
设计阶段：方案设计、初步设计、施工图设计	Design Phase：Concept Design, Development Design, Construction Documents Design
建设单位：映秀镇人民政府	Client：Yinxiu Town People's Government, Wenchuan
基地面积：14 300m^2	Site Area：14,300m^2
建筑面积：21 010m^2	Floor Area：21,010m^2
主要结构：钢筋混凝土框架结构	Main Structure：Reinforced Concrete Frame Structure
主要用途：住宅	Main Application：Residence

汶川映秀镇中心 15 号地块民房重建
Yinxiu Relief Residence Block15, Wenchuan

锦程川菜馆
震中饭

汶川映秀镇灾后恢复重建二台山项目
Yinxiu Relief Residence Block Ertaishar, Wenchuan

项目名称：汶川映秀镇灾后恢复重建二台山项目
建筑地点：汶川县映秀镇
设计时间：2008 年
竣工时间：2010 年
设计阶段：方案设计、初步设计、施工图设计
建设单位：映秀镇人民政府
基地面积：41 400m^2
建筑面积：21 175m^2
主要结构：钢筋混凝土框架结构
主要用途：住宅

Project: Yinxiu Relief Residence Block Ertaishar, Wenchuan
Construction Site: Yinxiu Town , Wenchuan
Design Period: 2008
Completion: 2010
Design Phase: Concept Design, Development Design, Construction Documents Design
Client: Yinxiu Town People's Government, Wenchuan
Site Area: 41,400m^2
Floor Area: 21,175m^2
Main Structure: Reinforced Concrete Frame Structure
Main Application: Residence

世博会城市最佳实践区景观工程
Landscape Project of UBPA 2010 EXPO, Shanghai

项目名称：世博会城市最佳实践区景观工程
建筑地点：上海
设计时间：2008 年
竣工时间：2010 年
设计阶段：方案设计、初步设计、施工图设计
建设单位：上海世博土地控股有限公司
基地面积：15.08hm^2
主要用途：景观

Project：Landscape Project of UBPA 2010 EXPO, Shanghai
Construction Site：Shanghai
Design Period：2008
Completion：2010
Design Phase：Concept Design, Development Design, Construction Documents Design
Client：Shanghai EXPO Land Holdings Co., Ltd.
Site Area：15.08hm^2
Main Application：Landscape

上海市南翔古檀园
Nanxiang Gutan Garden, Shanghai

项目名称：上海南翔古檀园
建设地点：上海市嘉定区
设计时间：2010 年
竣工时间：2012 年
设计阶段：方案设计、初步设计、施工图设计
建设单位：上海南翔老街建设发展有限公司
基地面积：8 700m²
主要用途：景观

Project：Nanxiang Gutan Garden, Shanghai
Project site：Jiading District, Shanghai
Design period：2010
Completion：2012
Design phase：Concept Design, Development Design, Construction Documents Design
Client：Shanghai Nanxiang Street Construction Development Co, Ltd.
Site area：8,700m²
Main Application：Landscape

盐城市廉政文化公园

Landscape Project of Anti-corruption Culture Park, Yancheng

项目名称：盐城市廉政文化公园
建筑地点：江苏省盐城市
设计时间：2012 年
竣工时间：2013 年
设计阶段：施工图设计
建设单位：盐城市市政公用投资有限公司
基地面积：24 450m^2
主要用途：社区公园

Project：Landscape Project of Anti-corruption Culture Park, Yancheng
Construction Site：Yancheng, Jiangsu
Design Period：2012
Completion：2013
Design Phase：Construction Design
Client：Yancheng City Municipal Public Investment Co., Ltd.
Site Area：24,450m^2
Main Application：Community Park

盐城市盐渎公园
Landscape Project of Yandu Park, Yancheng

项目名称：盐城市盐渎公园
建筑地点：江苏省盐城市
设计时间：2006 年
竣工时间：2007 年
设计阶段：方案设计、初步设计、施工图设计
建设单位：盐城市人民政府
基地面积：48.5hm^2
主要用途：综合性城市公园

Project：Landscape Project of Yandu Park, Yancheng
Construction Site：Yancheng, Jiangsu
Design Period：2006
Completion：2007
Design Phase：Concept Design, Development Design, Construction Documents Design
Client：Yancheng City People's Government
Site Area：48.5hm^2
Main Application：City Park

浦东竹园公园
Pudong Bamboo Park

项目名称：浦东竹园公园
建筑地点：上海市浦东新区
设计时间：2009 年
竣工时间：2009 年
设计阶段：施工图设计
建设单位：上海陆家嘴金融贸易区联合发展公司
基地面积：94 500m^2
主要用途：区域公园

Project：Pudong Bamboo Park
Construction Site：Pudong New District, Shanghai
Design Period：2009
Completion：2009
Design Phase：Construction Design
Client：Shanghai Lujiazui Finance and Trade United Development Company
Site Area：94,500m^2
Main Application：Regional Park

四川眉山东坡水月城市湿地公园

Meishan, Dongpo "Water Moon" Urban Wetland Park, Sichuan

项目名称：四川眉山东坡水月城市湿地公园
建筑地点：四川省眉山市
设计时间：2012 年
竣工时间：2014 年
设计阶段：修建性详细规划、方案设计
合作单位：上海比特景观建筑设计有限公司
建设单位：眉山市宏大建设投资有限责任公司
基地面积：69.48hm^2
主要用途：城市公园

Project：Meishan, Dongpo "Water Moon" Urban Wetland Park, Sichuan
Construction Site：Meishan, Sichuan
Design Period：2012
Completion：2014
Design Phase：Detailed Planning，Concept Design
Partnership：Shanghai BIT Landscape Architecture Design Co.,Ltd.
Client：Meishan Hongda Construction Investment Co.,Ltd.
Site Area：69.48hm^2
Main Application：City Park

遂宁体育中心景观工程

Landscape Project of Suining Sports Center, Sichuan

项目名称：遂宁体育中心景观工程
建筑地点：四川省遂宁市
设计时间：2010 年
竣工时间：2014 年
设计阶段：施工图设计
建设单位：遂宁市河东开发建设投资有限公司
基地面积：127 524.5m^2
主要用途：景观

Project：Landscape Project of Suining Sports Center, Sichuan
Construction Site：Suining, Sichuan
Design Period：2010
Completion：2014
Design Phase：Construction Design
Client：Suining River East Development and Construction Investment Co., Ltd.
Site Area：127,524.5m^2
Main Application：Landscape

崧泽博物馆景观工程
Landscape Project of Songze Museum, Shanghai

项目名称：崧泽博物馆景观工程
建筑地点：上海市青浦区
设计时间：2012 年
竣工时间：2014 年
设计阶段：施工图设计
建设单位：上海市文化广播影视管理局
基地面积：13 621m^2
主要用途：景观

Project：Landscape Project of Songze Museum, Shanghai
Construction site：Qingpu Distric Shanghai
Design Period：2012
Completion：2014
Design Phase：Construction Design
Client：Shanghai Municipal Bureau of Cultural Broadcasting and Television
Site Area：13,621m^2
Main Application：Landscape

项目名称：中国宣酒园区景观工程
建筑地点：安徽省宣城市
设计时间：2011—2012 年
竣工时间：2014 年
设计阶段：修建性详细规划
建设单位：安徽宣酒集团服务有限公司
基地面积：65.46hm^2
主要用途：景观

Project：Landscape Project of Xuan Wine Industry Culture Park, Xuancheng
Construction Site：Xuancheng, Anhui
Design Period：2011-2012
Completion：2014
Design Phase：Detailed Planning
Client：Anhui Xuan Wine Group Limited by Share Ltd.
Site Area：65.46hm^2
Main Application：Landscape

中国宣酒园区景观工程
Landscape Project of Xuan Wine Industry Culture Park, Xuancheng

中国宣酒文化园区纪叟雕塑实景

昆明市翠湖公园及环湖景观带提升与整治工程
Landscape Reform Project of Cuihu Park & Lakeside, Kunming

项目名称：昆明市翠湖公园及环湖景观带提升与整治工程
建筑地点：云南省昆明市
设计时间：2012 年
竣工时间：2012 年
设计阶段：初步设计
建设单位：昆明市五华区国有资产投资经营管理有限公司
基地面积：21.2 hm^2
主要用途：景观

Project：Landscape Reform Project of CuiHu Park & Lakeside, Kunming
Construction Site：Kunming, Yunnan
Design Period：2012
Completion：2012
Design Phase：Development Design
Client：Wuhua District State Owned Assets Investment Management Co., Ltd.
Site Area：21.2 hm^2
Main Application：Landscape

TJUADI 同济都市建筑

定聘设计人员
Part-time Specialists

佘 寅 SHE Yin
宋德萱 SONG Dexuan
孙澄宇 SUN Chengyu
孙光临 SUN Guan
孙彤宇 SUN Tongyu
汤朔宁 TANG Shuoning
童 明 TONG Ming
涂慧君 Huijun
王伯伟 WANG Bowei
王红军 WANG Hongjun
王桢栋 WANG Zhendong
王志军 WANG Zhijun
王 珂 WANG Ke
魏 崴 WEI Wei
吴长福 WU Changfu
吴庐生 heng
伍 江 WU Jiang
谢振宇 XIE Zhenyu
徐 风 XU Feng
徐 甘 XU Gan
徐磊青 XU Leiqing
颜宏亮 YAN Hongliang
杨 峰 YANG Feng
姚 栋 YAO Dong
俞 泳 YU Yong
袁 烽 YUAN Feng
张德顺 ZHANG Deshun
张 凡 ZHANG Fan
张建龙 ZHANG Jianlong
张 恺 ZHANG Kai
章 明 ZHANG Ming
张 鹏 ZHANG Peng
赵秀恒 ZHAO Xiuheng
郑时龄 ZHE Shiling
支文军 ZHI Wenjun
周 ZHOU Jian
周 健 ZHOU Jian
周向频 ZHOU Xiangpin
周友超 ZHOU Youchao
庄 宇 ZHUANG Yu

TJUADI 同济都市建筑

设计与管理人员
Full-time Specialists and Administration Staff

沈复宁 SHEN Funing
孙 璐 SUN Lu
王明充 WANG Mingchong
王 晔 WANG Ye
吴 沛 WU Pei
吴雨航 WU Yuhang
巫 蕊 WU Rui
奚凤新 XI Fengxin
徐 政 XU Zheng
叶文祺 YE Wenqi
余雪悦 YU Xueyue
闫红丽 YAN Hongli
张 瑾 ZHANG Jin
张 琪 ZHANG Qi
张亭康 ZHANG Tingkang
张文娟 ZHANG Wenjuan
张 毅 ZHANG Yi
张爱萍 ZHANG Aiping
张子岩 ZHANG Ziyan
赵英亓 ZHAO Yinqi
朱 赟 ZHU Yun
卓 宇 ZHUO Yu
张泽震 ZHANG Zezhen

TJUADI 同济都市建筑

活动掠影 Glimpse

各方支持与总结交流 Outside Resources and Communication

援建设计与专业调研 Post-disaster Reconstruction and Survey

学术活动与对外合作 Academic Activities and Inter-firm Collaboration

质量管理与职业培训 Quality Management and Professional Training

创作环境与现场服务 Working Environment and On-site Services

党群工作与企业文化 Party-people Connection and Enterprise Culture

图书在版编目（CIP）数据

同济都市建筑十年：2005－2015 / 同济大学建筑设计研究院（集团）有限公司都市建筑设计分院著. -- 上海：同济大学出版社，2015.11

ISBN 978-7-5608-6027-5

Ⅰ. ①同… Ⅱ. ①同… Ⅲ. ①建筑设计－作品集－中国－现代 Ⅳ. ①TU206

中国版本图书馆CIP数据核字（2015）第232166号

同济都市建筑十年：2005-2015

同济大学建筑设计研究院（集团）有限公司都市建筑设计院　著

责任编辑　荆　华　　责任校对　徐春莲　　装帧设计　朱丹天

出版发行　同济大学出版社 www.tongjipress.com.cn
（地址：上海四平路 1239 号　邮编：200092　电话：021-65985622）
经　　销　全国各地新华书店
印　　刷　上海安兴汇东纸业有限公司
开　　本　889mm×1194mm　1/16
印　　张　16
字　　数　512 000
版　　次　2015 年 11 月第 1 版　　2015 年 11 月第 1 次印刷
书　　号　ISBN 978-7-5608-6027-5
定　　价　280.00 元